SCOTTISH CERTIFICATE OF EDUCATION

Higher
MATHEMATICS

The Scottish Certificate of Education Examination Papers
are reprinted by special permission of
THE SCOTTISH EXAMINATION BOARD

Note: The answers to the questions do not emanate from the Board

ISBN 0 7169 9227 2
© *Robert Gibson & Sons, Glasgow, Ltd., 1996*

ROBERT GIBSON · Publisher
17 Fitzroy Place, Glasgow, G3 7SF.

MATHEMATICS   (Revised)

Higher Grade — PAPER I — Time: 2 hours
Higher Grade — PAPER II — Time: 2½ hours

INSTRUCTIONS TO CANDIDATES

READ CAREFULLY

1. Full credit will be given only where the solutions contains appropriate working.
2. Calculators may be used.
3. Answers obtained by readings from scale drawings will not receive any credit.

FORMULAE LIST

The equation $x^2 + y^2 + 2gx + 2fy + c = 0$ represents a circle centre $(-g, -f)$ and radius $\sqrt{(g^2 + f^2 - c)}$.

The equation $(x - a)^2 + (y - b)^2 = r^2$ represents a circle centre (a, b) and radius r.

Scalar Product: $a.b = |a||b| \cos \theta$ where θ is the angle between a and b

OR

$a.b = a_1 b_1 + a_2 b_2 + a_3 b_3$ where $a = \begin{pmatrix} a_1 \\ a_2 \\ a_3 \end{pmatrix}$ and $b = \begin{pmatrix} b_1 \\ b_2 \\ b_3 \end{pmatrix}$

Trigonometric formulae:
$$\sin (A \pm B) = \sin A \cos B \pm \cos A \sin B$$
$$\cos (A \pm B) = \cos A \cos B \mp \sin A \sin B$$
$$\cos 2A = \cos^2 A - \sin^2 A$$
$$= 2\cos^2 A - 1$$
$$= 1 - 2\sin^2 A$$
$$\sin 2A = 2\sin A \cos A$$

Table of standard derivatives:

$f(x)$	$f'(x)$
$\sin ax$	$a \cos ax$
$\cos ax$	$-a \sin ax$

Table of standard integrals:

$f(x)$	$\int f(x)dx$
$\sin ax$	$\dfrac{-1}{a} \cos ax + C$
$\cos ax$	$\dfrac{1}{a} \sin ax + C$

MATHEMATICS (REVISED)

Higher Grade—PAPER I
All questions should be attempted.

Marks

Find p if $(x + 3)$ is a factor of $x^3 - x^2 + px + 15$. **(3)**

Find the equation of the tangent to the curve $y = 4x^3 - 2$ at the point where $x = -1$. **(4)**

A circle passes through A$(-2, 3)$ and B$(4, -1)$.

Find the equation of the diameter which is perpendicular to the chord AB.

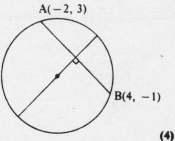

(4)

Show that P(2, 2, 3), Q(4, 4, 1) and R(5, 5, 0) are collinear and find the ratio in which Q divides PR. **(4)**

A cuboidal crystal is placed relative to the coordinate axes as shown opposite.

(a) Write down $\overrightarrow{BC}$ in component form.

(b) Calculate $|\overrightarrow{BC}|$.

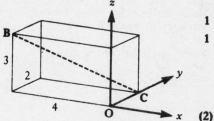

(2)

Evaluate $\displaystyle\int_1^2 (3x^2 + 4)\,dx$ and draw a sketch to illustrate the area represented by this integral. **(5)**

A bakery firm makes gingerbread men each 14 cm high with a circular "head" and "body".

The equation of the "body" is
$x^2 + y^2 - 10x - 12y + 45 = 0$
and the line of centres is parallel to the y-axis.

Find the equation of the "head".

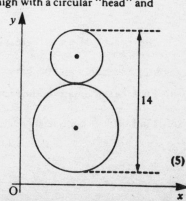

(5)

3

8. For all points on a curve $y = f(x)$, $f'(x) = 1 - 2x$.

   If the curve passes through the point (2, 1), find the equation of the curve.   (3

9. Given that $\cos D = \dfrac{2}{\sqrt{5}}$ and $0 < D < \dfrac{\pi}{2}$, find the exact values of $\sin D$ and $\cos 2D$.   (4

10. The diagram opposite shows the graph of a sine function from $0°$ to $90°$.

    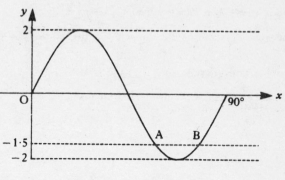

    (a) State the equation of the graph.   2

    (b) The line with equation $y = -1.5$ intersects the curve at A and B.

    Find the coordinates of A and B.   3

    (5

11. The diagram opposite shows a sketch of the cubic function f with stationary points at (0, 0) and (2, 4).

    Sketch the graph of the derived function f'.

    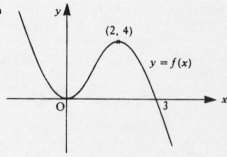

    (3)

12. The vector $ai + bj + k$ is perpendicular to both the vectors $i - j + k$ and $-2i + j + k$.

    Find the values of a and b.   (3)

13. (a) Find the coordinates of the points of intersection of the curves with equations
    $$y = 2x^2 \text{ and } y = 4 - 2x^2 .$$ 2

    (b) Find the area completely enclosed between these two curves.   3

    (5)

Marks

4. As shown in the diagram opposite, a set of
   experimental results gives a straight line
   graph when $\log_{10}y$ is plotted against $\log_{10}x$.
   The straight line passes through $(0, 1)$
   and has a gradient of 2.
   Express y in terms of x.

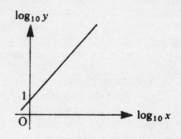

(6)

5. Solve the equation $2\cos^2 x = \tfrac{1}{2}$ for $0 \leqslant x \leqslant \pi$. **(3)**

6. For what values of x is the function $f(x) = \tfrac{1}{3}x^3 - 2x^2 - 5x - 4$ increasing? **(5)**

7. Make a copy of this graph of $y = \log_{10}x$.
   On your copy, sketch the graph of
   $$y = \log_{10}(x - 2).$$

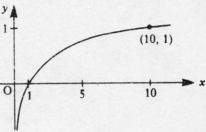

(3)

8. Show that the roots of the equation $(k - 2)x^2 - (3k - 2)x + 2k = 0$ are real. **(4)**

9. If $f(x) = \cos^2 x - \dfrac{2}{3x^2}$, find $f'(x)$. **(4)**

20. The right-angled triangle OAB with sides
    of length 3 cm, 4 cm and 5 cm is placed
    with one vertex at the origin O as
    shown in the diagram.
    A circle centre C and diameter RO
    of length 13 cm is drawn and passes
    through O and B.
    What is the gradient of the line RO?

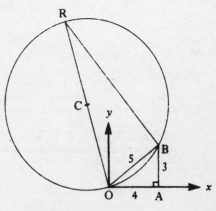

(5)

MATHEMATICS (REVISED)

Higher Grade—PAPER II

All questions should be attempted.

Marks

1. A function f is defined by the formula $f(x) = (x - 1)^2(x + 2)$ where $x \in \mathbf{R}$.
   - (a) Find the coordinates of the points where the curve with equation $y = f(x)$ crosses the x- and y-axes. 3
   - (b) Find the stationary points of this curve $y = f(x)$ and determine their nature. 7
   - (c) Sketch the curve $y = f(x)$. 2

   (12)

2. P, Q and R have coordinates $(1, -2)$, $(6, 3)$ and $(9, 14)$ respectively and are three vertices of a kite PQRS.
   - (a) Find the equations of the diagonals of this kite and the coordinates of the point where they intersect. 7
   - (b) Find the coordinates of the fourth vertex S. 2

   (9)

3. The extract below is taken from the "OIL RIG NEWS".

   ## RARE ILLNESS STRIKES RIG

   ### Storm prevents delivery of medicine

   By noon on Tuesday 20 December 1988 50 of our oil rig personnel were laid low by a mystery illness.

   Our resident medical officer is expressing concern because the number of personnel affected is increasing each day by 8% of the previous day's total.

   - (a) If the daily rate of increase remained at 8% of the previous day's total, how many personnel were affected by noon on Sunday 25 December 1988? 3
   - (b) An improvement in the weather conditions allowed a team of medics to fly out to the rig on the morning of Tuesday 27 December 1988.

     At noon on that Tuesday, all personnel were inoculated and no new cases of the illness arose. Within the next 24 hours, 21% of those who had been affected had recovered.

     If the daily rate of recovery of 21% of the previous day's total was maintained, how many personnel were still affected by the illness at noon on Saturday 31 December 1988? 5

   (8)

4. This is an extract from a newspaper article concerning the new entrance to the Louvre Museum.

President Mitterrand yesterday unveiled a controversial giant glass pyramid destined to be the new entrance to the Louvre Museum in Paris.

(*a*) Relative to the mutually perpendicular axes O*x*, O*y* and O*z*, the front face of this pyramid is represented by triangle ABC, where A is the point with coordinates (9, 9, 24), B is the point (27, 3, 0), C is the point (3, 27, 0) and M is the mid-point of AC, as shown in the diagram below.

Find the coordinates of G which divides BM in the ratio 2 : 1.   **3**

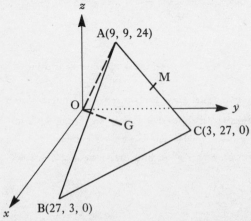

(*b*) Support girders are to be erected from O to G and from O to A.

Calculate the size of the angle between the girders.   **5**

(8)

5. (a) Show that $2\cos(x + 30)° - \sin x°$ can be written as $\sqrt{3}\cos x° - 2\sin x°$.   **3**

(b) Express $\sqrt{3}\cos x° - 2\sin x°$ in the form $k\cos(x + \alpha)°$ where $k > 0$ and $0 < \alpha < 360$ and find the values of k and α.   **4**

(c) Hence, or otherwise, solve the equation $2\cos(x + 30)° = \sin x° + 1$, $0 \leqslant x \leqslant 360$.   **3**

(10)

6. (a) The function f is defined by $f(x) = x^3 - 2x^2 - 5x + 6$.
The function g is defined by $g(x) = x - 1$.
Show that $f(g(x)) = x^3 - 5x^2 + 2x + 8$.   **4**

(b) Factorise fully $f(g(x))$.   **3**

(c) The function k is such that $k(x) = \dfrac{1}{f(g(x))}$.

For what values of x is the function k not defined?   **2**

(9)

7. The diagram shows two curves with equations $y = x^2$ and $y^2 = x$.

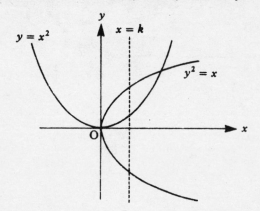

The area completely enclosed between these two curves is divided in half by the line with equation $x = k$.

(a) Represent these two equal areas by two separate integrals each involving k.   **6**

(b) Equate the integrals and show that k is given by the equation
$$2k^3 - 4k^{\frac{3}{2}} + 1 = 0 .$$
  4

(c) Use the substitution p^2 for k^3 to find the value of k.   **4**

(14)

8. A sports club awards trophies in the form of paperweights bearing the club crest.

Diagram 1 shows the front view of one of these paperweights.

Each is made from two different types of glass. The two circles are concentric and the base line is a tangent to the inner circle.

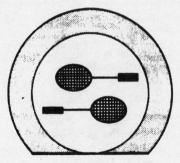

Diagram 1

(a) Relative to x, y coordinate axes, the equation of the outer circle is
$$x^2 + y^2 - 8x + 2y - 19 = 0$$
and the equation of the base line is $y = -6$.

Show that the equation of the inner circle is $x^2 + y^2 - 8x + 2y - 8 = 0$.   **4**

(b) An alternative form of the paperweight is made by cutting off a piece of glass from the original design along a second line with equation $3x - 4y + 9 = 0$ as shown in Diagram 2.

Show that this line is a tangent to the inner circle and state the coordinates of the point of contact.

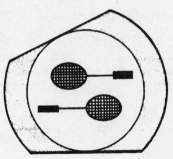

   7

Diagram 2

                                               (11)

9. Polynomial equations often have roots which are not whole numbers.

One method of estimating the roots of such equations is to make repeated use of the following:

> If $x = p$ is an estimate of a root of the equation $f(x) = 0$, then $x = q$ will be a closer estimate where
> $$q = p - \frac{f(p)}{f'(p)}$$

EXAMPLE

One of the roots of the equation $x^2 - 2x - 5 = 0$ is known to lie between 3 and 4. The above rule can be used to estimate this root to one decimal place as follows:

$$f(x) = x^2 - 2x - 5 \quad \text{so } f'(x) = 2x - 2$$

Choose $p = 3$ **(1st estimate)**

then $q = 3 - \frac{f(3)}{f'(3)} = 3 - \frac{(-2)}{4} = \mathbf{3 \cdot 5}$

> Hence $x = 3 \cdot 5$ is a closer estimate to the root than $x = 3$

Choose $p = 3 \cdot 5$ **(2nd estimate)**

then $q = 3 \cdot 5 - \frac{f(3 \cdot 5)}{f'(3 \cdot 5)} = 3 \cdot 5 - \frac{0 \cdot 25}{5} = \mathbf{3 \cdot 45}$

> Hence $x = 3 \cdot 45$ is a closer estimate to the root than $x = 3 \cdot 5$

Choose $p = 3 \cdot 45$ **(3rd estimate)**

then $q = 3 \cdot 45 - \frac{f(3 \cdot 45)}{f'(3 \cdot 45)} = 3 \cdot 45 - \frac{0 \cdot 0025}{4 \cdot 9} = \mathbf{3 \cdot 449}$

> Hence $x = 3 \cdot 449$ is a closer estimate to the root than $x = 3 \cdot 45$

CONCLUSION The root, correct to one decimal place, is $x = \mathbf{3 \cdot 4}$

(a) Show that the equation $x^3 - 2x^2 + 6x - 4 = 0$ has a root between 0 and 1. **3**

(b) Use the method described above to find this root correct to one decimal place. **6**

(9)

10. The Water Board of a local authority discovered it was able to represent the approximate amount of water $W(t)$, in millions of gallons, stored in a reservoir t months after 1st May 1988 by the formula

$$W(t) = 1 \cdot 1 - \sin \frac{\pi t}{6}$$

The board then predicted that under normal conditions this formula would apply for three years.

(a) Draw and label sketches of the graphs of $y = \sin \frac{\pi t}{6}$ and $y = -\sin \frac{\pi t}{6}$, for $0 \leqslant t \leqslant 36$ on the same diagram.

4

(b) On a separate diagram and using the same scale on the t-axis as you used in part (a), draw a sketch of the graph of $W(t) = 1 \cdot 1 - \sin \frac{\pi t}{6}$.

3

(c) On the 1st April 1990 a serious fire required an extra $\frac{1}{4}$ million gallons of water from the reservoir to bring the fire under control.

Assuming that the previous trend continues from the new lower level, when will the reservoir run dry if water rationing is not imposed?

3

(10)

MATHEMATICS (REVISED)

Higher Grade—PAPER I

All questions should be attempted

Marks

1. Find the equation of the line through the point $(3, -5)$ which is parallel to the line with equation $3x + 2y - 5 = 0$.

   (2)

2. The points A and B have coordinates (a, a^2) and $(2b, 4b^2)$ respectively. Determine the gradient of AB in its simplest form.

   (2)

3. Show that the vectors $a = 2i + 3j - k$ and $b = 3i - j + 3k$ are perpendicular.

   (3)

4. Diagram 1 shows part of the graph of $y = ke^{0.5x}$.

   (a) Find the value of k.

   1

   (b) The line with equation $x = 1$ intersects the graph at P.

   Find the coordinates of the point P.

   2

   (3)

Diagram 1

5. Find the equation of the tangent to the curve $y = 3x^2 + 2$ at the point where $x = 1$.

   (4)

6. When $f(x) = 2x^4 - x^3 + px^2 + qx + 12$ is divided by $(x - 2)$, the remainder is 114. One factor of $f(x)$ is $(x + 1)$.

   Find the values of p and q.

   (5)

7. (a) Show that the points L$(-5, 6, -5)$, M$(7, -2, -1)$ and N$(10, -4, 0)$ are collinear.

   4

   (b) Find the ratio in which M divides LN.

   1

   (5)

8. Find the equation of the tangent at the point $(3, 1)$ on the circle $x^2 + y^2 - 4x + 6y - 4 = 0$.

   (5)

Marks

9. Diagram 2 shows the graph of
$y = f(x)$, where $-2 \leqslant x \leqslant 3$.

   On separate diagrams, sketch the
   graphs of:

   (a) $y = -f(x)$;
   (b) $y = f'(x)$.

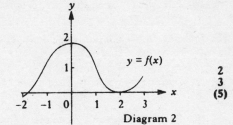

Diagram 2

2
3
(5)

10. A curve with equation $y = f(x)$ passes through the point $(2, -1)$ and is such
    that $f'(x) = 4x^3 - 1$.

    Express $f(x)$ in terms of x.

(5)

11. On the day of his thirteenth birthday, a boy is given a sum of money to
    invest and instructions not to withdraw any money until after his
    eighteenth birthday. The money is invested and compound interest of 9%
    per annum is added each following birthday. By what percentage will the
    investment have increased when he withdraws his money just after his
    eighteenth birthday?

(4)

12. Given that $\sin A = \frac{3}{4}$, where $0 < A < \frac{\pi}{2}$, find the **exact** value of $\sin 2A$.

(3)

13. Given that $f(x) = 5(7 - 2x)^3$, find the value of $f'(4)$.

(4)

14. Diagram 3 is a rough sketch of the curves $y = f(x)$, $y = g(x)$ and $y = h(x)$.
    A is $(-4, -12)$, B is $(-2, 0)$, C is $(-1, -3)$, D is $(1, 3)$, E is $(2, 0)$ and F is $(4, 12)$.

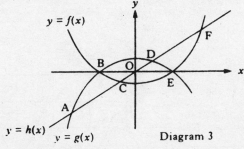

Diagram 3

    State the range of values of x for which:

    (a) $f(x) \leqslant g(x)$;
    (b) $h(x) < g(x) < f(x)$.

1
2
(3)

15. (*a*) Express $7 - 2x - x^2$ in the form $a - (x + b)^2$ and write down the values of *a* and *b*. **2**

(*b*) State the maximum value of $7 - 2x - x^2$ and justify your answer. **2**
(4)

16. (*a*) Find the value of $\displaystyle\int_1^2 (4 - x^2)dx$. **3**

(*b*) Sketch a graph and shade the area represented by the integral in part (*a*). **2**
(5)

17. Diagram 4 shows a right-angled isosceles triangle whose sides represent the vectors **a**, **b** and **c**. The two equal sides have length 2 units.

Find the value of **b** . (**a** + **b** + **c**). **(5)**

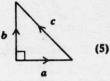

Diagram 4

18. Given that *k* is a real number, show that the roots of the equation $kx^2 + 3x + 3 = k$ are always real numbers. **(5)**

19. Diagram 5 illustrates three functions *f*, *g* and *h*. The functions *f* and *g* are defined by

$$f(x) = 2x + 5$$
$$g(x) = x^2 - 3.$$

The function *h* is such that whenever $f(p) = q$ and $g(q) = r$, then $h(p) = r$.

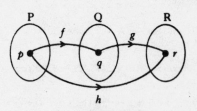

Diagram 5

(*a*) If $q = 7$, find the values of *p* and *r*. **2**
(*b*) Find a formula for $h(x)$, in terms of *x*. **2**
(4)

20. Diagram 6 shows two curves

$y = \cos 2x°$ and
$y = 1 + \sin x°$,
where $0 \leqslant x \leqslant 360$.

Find the *x*-coordinate of the point of intersection at A.

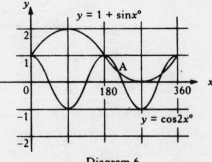

Diagram 6 **(4)**

14

MATHEMATICS (REVISED)

Higher Grade—PAPER II

All questions should be attempted

Marks

1. (a) Diagram 1 shows a part of the curve with equation $y = 2x^2(x - 3)$. Find the coordinates of the stationary points on the graph and determine their nature.

    4

    (b) State the range of values of k for which $y = k$ intersects the graph in three distinct points.

    2
    (6)

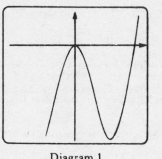

Diagram 1

2. (a) In Diagram 2, A is the point $(-1, 1)$, B is $(3, 3)$ and C is $(6, 2)$. The perpendicular bisector of AB has equation $y + 2x = 4$. Find the equation of the perpendicular bisector of BC.

    4

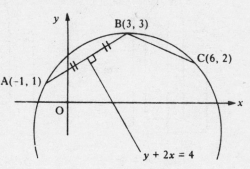

Diagram 2

    (b) Find the centre and the equation of the circle which passes through A, B and C.

    6
    (10)

3. Diagram 3 shows an isosceles triangle PQR in which PR = QR and angle PQR = $x°$.

    (a) Show that $\dfrac{\sin x°}{p} = \dfrac{\sin 2x°}{r}$.

    4

    (b) (i) State the value of $x°$ when $p = r$.

    1

    (ii) Using the fact that $p = r$, solve the equation in (a) above, to justify your stated value of $x°$.

    4
    (9)

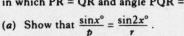
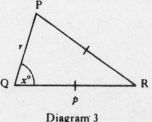

Diagram 3

4. (*a*) On the same diagram, sketch the graphs of $y = \log_{10}x$ and $y = 2 - x$ where $0 < x < 5$.

Write down an approximation for the x-coordinate of the point of intersection.   **3**

(*b*) Find the value of this x-coordinate, correct to 2 decimal places.   **3**

(6)

5. Diagram 4 shows a Christmas tree decoration which is made of coloured glass rods in the shape of a square-based prism topped by a square pyramid. Diagram 5 shows the decoration relative to origin O and rectangular coordinate axes OX, OY and OZ.

The vertex F has position vector $\begin{pmatrix} 2 \\ 2 \\ -7 \end{pmatrix}$ and the vertex V has position

vector $\begin{pmatrix} 1 \\ 1 \\ 3 \end{pmatrix}$.

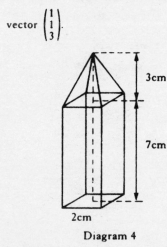

Diagram 4

Diagram 5

(*a*) Find:

(i) the components of the vectors represented by $\overrightarrow{VF}$ and $\overrightarrow{VE}$;   **3**

(ii) the size of the angle EVF.   **4**

(*b*) To make the decoration more attractive, triangular sheets of coloured glass VEF and VDG are added to it.

Calculate the area of the glass triangle VEF.   **3**

(10)

16

6. There is a rule known as the Product Rule which is used, as shown below, to differentiate any product of two functions of the same variable.

THE PRODUCT RULE

If $P(x) = f(x) \cdot g(x)$, then $P'(x) = f'(x) \cdot g(x) + f(x) \cdot g'(x)$

EXAMPLE: Find the derivative of $P(x) = x^2 \sin x$

$P(x) = x^2 \cdot \sin x$ Choose $f(x) = x^2$ and $g(x) = \sin x$
                                      hence $f'(x) = 2x$ and $g'(x) = \cos x$

$P'(x) = 2x \cdot \sin x + x^2 \cdot \cos x$

$P'(x) = 2x\sin x + x^2 \cos x$

Use the Product Rule to find the derivative of $P(x) = x^3 \cos x$.   **(5)**

7. (*a*) A tractor tyre is inflated to a pressure of 50 units.
Twenty-four hours later the pressure has dropped to 10 units.

If the pressure, P_t units, after t hours is given by the formula $P_t = P_0 e^{-kt}$, find the value of k, to three decimal places.   **4**

(*b*) The tyre manufacturer advises that serious damage to the tyre will result if it is used when the pressure drops below 30 units.

If the farmer inflates the tyre to 50 units and drives the tractor for 4 hours, can the tractor be driven further without inflating the tyre and without risking serious damage to the tyre?   **5**
   (9)

8. The displacement, *d* units, of a wave after *t* seconds, is given by the formula
$$d = \cos 20t° + \sqrt{3}\sin 20t°.$$

(*a*) Express *d* in the form $k\cos(20t - \alpha)°$, where $k > 0$ and $0 \leqslant \alpha \leqslant 360$.   **4**

(*b*) Sketch the graph of *d* for $0 \leqslant t \leqslant 18$.   **4**

(*c*) Find, correct to 1 decimal place, the values of t, $0 \leqslant t \leqslant 18$, for which the displacement is 1·5 units.   **4**
   (12)

9. (*a*) At 12 noon a hospital patient is given a pill containing 50 units of antibiotic.
By 1p.m. the number of units in the patient's body has dropped by 12%.
By 2p.m. a further 12% of the units remaining in the body at 1p.m. is lost.

If this fall-off rate is maintained, find the number of units of antibiotic remaining at 6p.m. **3**

(*b*) A doctor considers prescribing a course of treatment which involves a patient taking one of these pills every 6 hours over a long period of time.
The doctor knows that more than 100 units of this antibiotic in the body is regarded as too dangerous.

Should the doctor prescribe this course of treatment or not?
Give reasons for your answer. **6**
(9)

10. Diagram 6 shows a rectangular sheet of transparent plastic moulded into a parabolic shape and pegged to the ground to form a cover for growing plants. Triangular metal frames are placed over the cover to support it and prevent it blowing away in the wind.

Diagram 7 shows an end view of the cover and the triangular frame related to the origin O and axes O*x* and O*y*. (All dimensions are given in centimetres.)

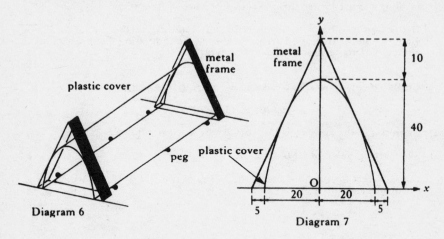

Diagram 6

Diagram 7

(*a*) Show that the equation of the parabolic end is $y = 40 - \dfrac{x^2}{10}$, $-20 \leqslant x \leqslant 20$. **4**

(*b*) Show that the triangular frame touches the cover without disturbing the parabolic shape. **7**
(11)

18

11. Diagram 8 shows a proposed replacement of the old platform canopy at the local railway station by a new parabolic canopy, while keeping the original pillars.

If OR and OP are taken as the x- and y-axes and Q has coordinates $(1, 1)$, then OQ has equation $y = \sqrt{x}$ and PQ is the tangent at Q to the parabola.

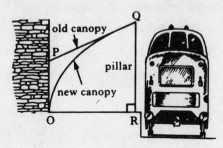

Diagram 8

The planners have received an objection that there is a reduction of more than 10% in the space under the canopy and wish to compare the two canopies.

(*a*) Find the equation of the tangent PQ and the coordinates of P.   5

(*b*) Find the area of the trapezium OPQR.   2

(*c*) Find the area under the parabola OQ.   3

(*d*) Comment on the objection received.   3
                                                                              (13)

[END OF QUESTION PAPER]

SCOTTISH CERTIFICATE OF EDUCATION

MATHEMATICS (REVISED)

Higher Grade—PAPER I

Monday, 11th May—9.30 a.m. to 11.30 a.m.

All questions should be attempted

Marks

1. Find the equation of the tangent to the curve with equation $y = 5x^3 - 6x^2$ at the point where $x = 1$. **(4)**

2. In the diagram, A is the point (7, 0),
   B is (−3, −2) and C (−1, 8).
   The median CE and the altitude BD intersect at J.
   (*a*) Find the equations of CE and BD.

   (*b*) Find the coordinates of J.

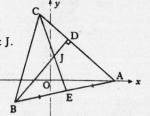

(8)

3. Find k if $x - 2$ is a factor of $x^3 + kx^2 - 4x - 12$. **(3)**

4. A curve for which $\dfrac{dy}{dx} = 3x^2 + 1$ passes through the point (−1, 2).

   Express y in terms of x. **(4)**

5. Find, correct to one decimal place, the value of x between 180 and 270 which satisfies the equation $3\cos(2x - 40)° - 1 = 0$. **(5)**

6. On a suitable set of real numbers, functions f and g are defined by

$$f(x) = \frac{1}{x + 2} \quad \text{and} \quad g(x) = \frac{1}{x} - 2.$$

   Find $f(g(x))$ in its simplest form. **(3)**

7. (*a*) Express $\sin x° - 3\cos x°$ in the form $k \sin(x - \alpha)°$ where $k > 0$ and $0 \leqslant \alpha < 360$. Find the values of k and α.

   (*b*) Find the maximum value of $5 + \sin x° - 3\cos x°$ and state a value of x for which this maximum occurs. **(6)**

8. Evaluate $\displaystyle\int_1^9 \frac{x + 1}{\sqrt{x}}\, dx$. **(5)**

9.

An ancient Stone Circle has a processional pathway from the Heelstone to the centre of the Stone Circle. In the picture above, the Heelstone is on the left and the dotted line represents the processional pathway.

With suitable axes and using the Heelstone as the origin, the equation of the Stone Circle is $x^2 + y^2 - 8x - 6y + 21 = 0$.

Given that 1 unit represents 15 metres, calculate the distance in metres from the Heelstone to the nearest point on the edge of the Circle.   **(5)**

10. The sketch shows the graph of $y = f(x)$ for $-2 \leqslant x \leqslant 4$.

The function $g(x)$ has the line $x = 4$ as an axis of symmetry and $g(x) = f(x)$ for $-2 \leqslant x \leqslant 4$.

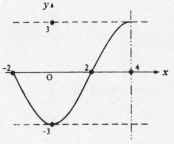

On separate sketches, indicate

(a) $y = g(x)$ for $-2 \leqslant x \leqslant 10$

(b) $y = -2g(x)$ for $0 \leqslant x \leqslant 8$.   **(4)**

11. Differentiate $2x^{\frac{3}{2}} + \sin^2 x$ with respect to x.   **(4)**

12. An incomplete sketch (not drawn to scale) of the graph of $y = \log_{10}(x + a)$ is shown. Find the value of a.   **(2)**

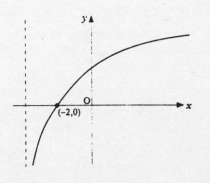

13. The diagram shows a kite OABC. A is the point (4,0) and B is the point (4,3).

Calculate the gradient of OC correct to two decimal places.

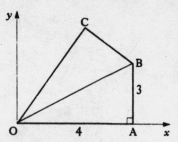

(3)

14. (*a*) Evaluate $\int_{0}^{\pi/2} \cos 2x \, dx$.

(*b*) Draw a sketch and explain your answer.

(5)

15.

An aircraft flying at a constant speed on a straight flight path takes 2 minutes to fly from A to B and 1 minute to fly from B to C. Relative to a suitable set of axes, A is the point (−1,3,4) and B is the point (3, 1,−2). Find the coordinates of the point C.

(3)

16. AB is a tangent at B to the circle with centre C and equation

$$(x-2)^2 + (y-2)^2 = 25.$$

The point A has coordinates (10,8).

Find the area of triangle ABC.

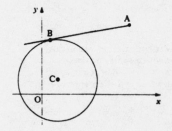

(5)

Marks

17. Calculate the least positive integer value of k so that the graph of $y = kx^2 - 8x + k$ does not cut or touch the x-axis.

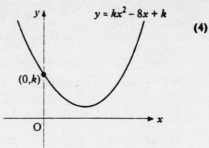

(4)

18. The diagram shows representatives of two vectors, a and b, inclined at an angle of 60°.

   If $|a| = 2$ and $|b| = 3$, evaluate $a.(a + b)$.

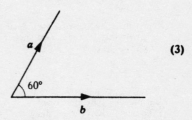

(3)

19. The line with equation $y = x$ is a tangent at the origin to the parabola with equation $y = f(x)$. The parabola has a maximum turning point at (a, b).

   Sketch the graph of $y = f'(x)$.

(4)

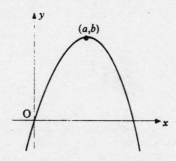

[END OF QUESTION PAPER]

23

MATHEMATICS (REVISED)

Higher Grade—PAPER II

Monday, 11th May—1.30 p.m. to 4.00 p.m.

All questions should be attempted

Marks

1. The diagram shows part of the graph of the curve with equation

   $f(x) = x^3 + x^2 - 16x - 16$.

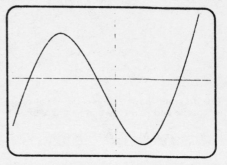

   (a) Factorise $f(x)$.

   3

   (b) Write down the coordinates of the four points where the curve crosses the x and y axes.

   2

   (c) Find the coordinates of the turning points and justify their nature.

   6

   (11)

2. Relative to a suitable set of coordinate axes with a scale of 1 unit to 2 kilometres, the positions of a transmitter mast, ship, aircraft and satellite dish are shown in the diagram below.

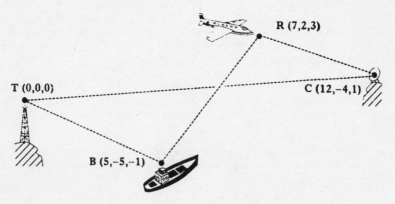

   The top T of the transmitter mast is the origin, the bridge B on the ship is the point $(5,-5,-1)$, the centre C of the dish on the top of a mountain is the point $(12,-4,1)$ and the reflector R on the aircraft is the point $(7,2,3)$.

   (a) Find the distance in kilometres from the bridge of the ship to the reflector on the aircraft.

   3

   (b) Three minutes earlier, the aircraft was at the point $M(-2,4,8\cdot5)$. Find the speed of the aircraft in kilometres per hour.

   2

   (c) Prove that the direction of the beam TC is perpendicular to the direction of the beam BR.

   3

   (d) Calculate the size of angle TCR.

   5

   (13)

3. Biologists calculate that, when the concentration of a particular chemical in a sea loch reaches 5 milligrams per litre (mg/l), the level of pollution endangers the life of the fish.

A factory wishes to release waste containing this chemical into the loch. It is claimed that the discharge will not endanger the fish.

The Local Authority is supplied with the following information:

---

1. The loch contains none of this chemical at present.

2. The factory manager has applied to discharge waste once per week which will result in an increase in concentration of 2·5 mg/l of the chemical in the loch.

3. The natural tidal action will remove 40% of the chemical from the loch every week.

(a) Show that this level of discharge would result in fish being endangered.   3

When this result is announced, the company agrees to install a cleaning process that reduces the concentration of chemical released into the loch by 30%.

(b) Show the calculations you would use to check this revised application.

Should the Local Authority grant permission?   5

                                                                                   (8)

4. (a) For a particular radioactive substance, the mass m (in grams) at time t (in years) is given by

$$m = m_0 e^{-0.02t}$$

where m_0 is the original mass.

If the original mass is 500 grams, find the mass after 10 years.   2

(b) The half-life of any material is the time taken for half of the mass to decay.

Find the half-life of this substance.   3

(c) Illustrate **all** of the above information on a graph.   3

                                                                                   (8)

5. The owners of a zoo intend to build a new aviary in the shape of a cuboid with a square floor. The volume of the aviary will be 500 m^3.

(*a*) If x metres is the length of one edge of the floor, show that the area A square metres of netting required is given by

$$A = x^2 + \frac{2000}{x}$$

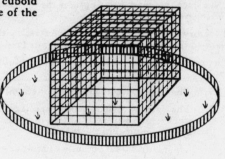

4

(*b*) Find the dimensions of the aviary to ensure that the cost of netting is minimised.

6

(10)

6. The *vector product*, $a \times b$, of $a = \begin{pmatrix} a_1 \\ a_2 \\ a_3 \end{pmatrix}$ and $b = \begin{pmatrix} b_1 \\ b_2 \\ b_3 \end{pmatrix}$ is defined by

$$a \times b = \begin{pmatrix} a_2 b_3 - a_3 b_2 \\ a_3 b_1 - a_1 b_3 \\ a_1 b_2 - a_2 b_1 \end{pmatrix}.$$

Example

When $a = \begin{pmatrix} 1 \\ 2 \\ 3 \end{pmatrix}$ and $b = \begin{pmatrix} -1 \\ 0 \\ 2 \end{pmatrix}$

then $a \times b = \begin{pmatrix} 2 \times 2 - 3 \times 0 \\ 3 \times (-1) - 1 \times 2 \\ 1 \times 0 - 2 \times (-1) \end{pmatrix} = \begin{pmatrix} 4 \\ -5 \\ 2 \end{pmatrix}.$

(*a*) If a and b are as shown in the diagram and $c = a \times b$, evaluate c.

3

(*b*) By considering $a.c$ and $b.c$, what can be concluded about c?

4

(7)

(2,1,1)

(3,−1,2)

(4,1,0)

b

a

Marks

7. (*a*) Solve the equation $3\sin 2x° = 2\sin x°$ for $0 \leqslant x \leqslant 360$. **4**

(*b*) The diagram below shows parts of the graphs of sine functions f and g.
State expressions for $f(x)$ and $g(x)$. **1**

(*c*) Use your answers to part (*a*) to find the coordinates of A and B. **2**

(*d*) Hence state the values of x in the interval $0 \leqslant x \leqslant 360$ for which $3\sin 2x° < 2\sin x°$. **3**

(10)

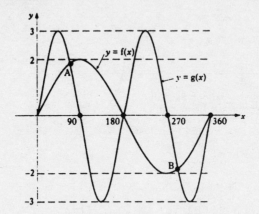

8. A ship is sailing due north at a constant speed.

When at position A, a lighthouse L is observed on a bearing of $a°$. One hour later, when the ship is at position B, the lighthouse is on a bearing of $b°$.

The shortest distance between the ship and the lighthouse during this hour was d miles.

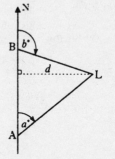

(*a*) Prove that $AB = \dfrac{d}{\tan a°} - \dfrac{d}{\tan b°}$. **2**

(*b*) Hence prove that $AB = \dfrac{d\sin(b-a)°}{\sin a° \sin b°}$. **3**

(*c*) Calculate the shortest distance from the ship to the lighthouse when the bearings $a°$ and $b°$ are 060° and 135° respectively and the constant speed of the ship is 14 miles per hour. **3**

(8)

BALLOON

GONDOLA

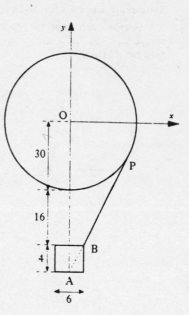

A spherical hot-air balloon has radius 30 feet. Cables join the balloon to the gondola which is cylindrical with diameter 6 feet and height 4 feet. The top of the gondola is 16 feet below the bottom of the balloon.

Coordinate axes are chosen as shown in the diagram. One of the cables is represented by PB and PBA is a straight line.

(a) Find the equation of the cable PB.   **3**

(b) State the equation of the circle representing the balloon.   **1**

(c) Prove that this cable is a tangent to the balloon and find the coordinates of the point P.   **5**

                                                                                (9)

10. An artist has been asked to design a window made from pieces of coloured glass with different shapes. To preserve a balance of colour, each shape must have the **same** area. Three of the shapes used are drawn below.

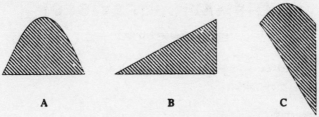

A   B   C

Relative to x,y-axes, the shapes are positioned as shown below. The artist drew the curves accurately by using the equation(s) shown in each diagram.

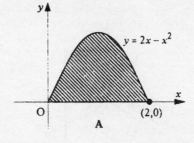

(*a*) Find the area shaded under $y = 2x - x^2$.

4

$y = 2x - x^2$

A

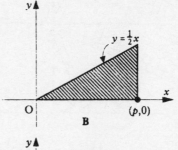

(*b*) Use the area found in part (*a*) to find the value of p.

2

$y = \frac{1}{2}x$

B

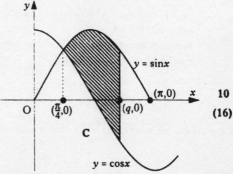

(*c*) Prove that q satisfies the equation $\cos q + \sin q = 0 \cdot 081$ and hence find the value of q to 2 significant figures.

10

(16)

$y = \sin x$

$y = \cos x$

C

[END OF QUESTION PAPER]

SCOTTISH CERTIFICATE OF EDUCATION

MATHEMATICS (REVISED)

Higher Grade—PAPER I

All questions should be attempted   *Marks*

1. A is the point $(-3,2,4)$ and B is $(-1,3,2)$. Find

    (*a*) the components of vector $\overrightarrow{AB}$;   **(1)**

    (*b*) the length of AB.   **(2)**

2. Relative to the axes shown and with an appropriate scale, Alex stands at the point $(-2,3)$ where Hartington Road meets Newport Road.

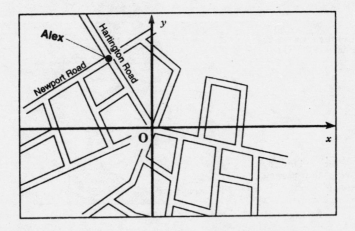

    (*a*) Find the equation of Newport Road which is perpendicular to
           Hartington Road.   **(3)**

    (*b*) Brenda is waiting for a bus at the point $(-5,1)$. Show that Brenda is
           standing on Newport Road.   **(1)**

3. Find the values of k for which the equation $2x^2 + 4x + k = 0$ has real roots.   **(2)**

4. Find the x-coordinate of each of the points on the curve
    $y = 2x^3 - 3x^2 - 12x + 20$ at which the tangent is parallel to the x-axis.   **(4)**

5. This diagram shows a computer-generated display of a game of noughts and crosses. Relative to the coordinate axes which have been added to the display, the "nought" at A is represented by a circle with equation

$(x-2)^2 + (y-2)^2 = 4$.

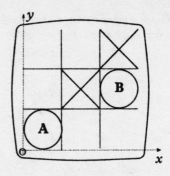

(a) Find the centre of the circle at B. **(3)**

(b) Find the equation of the circle at B. **(1)**

6. For acute angles P and Q, $\sin P = \frac{12}{13}$ and $\sin Q = \frac{3}{5}$.

Show that the **exact** value of $\sin(P+Q)$ is $\frac{63}{65}$. **(3)**

7. One root of the equation $2x^3 - 3x^2 + px + 30 = 0$ is -3.
Find the value of p and the other roots. **(4)**

8. The diagram shows the graph of $y = f(x)$.

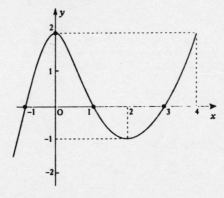

Sketch the graph of $y = 2 - f(x)$. **(3)**

9. Differentiate $4\sqrt{x} + 3\cos 2x$. **(4)**

10. The lines $y = 2x + 4$ and $x + y = 13$ make angles of $a°$ and $b°$ with the positive direction of the x-axis, as shown in the diagram.

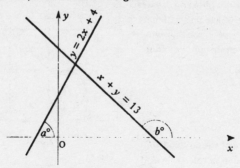

(*a*) Find the values of a and b. **(4)**

(*b*) **Hence** find the acute angle between the two given lines. **(1)**

11. The graphs of $y = f(x)$ and $y = g(x)$ intersect at the point A on the y-axis, as shown on the diagram.

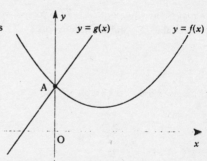

If $g(x) = 3x + 4$
and $f'(x) = 2x - 3$, find $f(x)$. **(4)**

12. The vectors *a, b* and *c* are defined as follows:

$$a = 2i - k, \quad b = i + 2j + k, \quad c = -j + k.$$

(*a*) Evaluate $a.b + a.c$. **(3)**

(*b*) From your answer to part (*a*), make a deduction about the vector $b + c$. **(2)**

32

13. $f(x) = 2x - 1$, $g(x) = 3 - 2x$ and $h(x) = \frac{1}{4}(5 - x)$.

    (*a*) Find a formula for $k(x)$ where $k(x) = f\big(g(x)\big)$.   **(2)**

    (*b*) Find a formula for $h\big(k(x)\big)$.   **(2)**

    (*c*) What is the connection between the functions h and k?   **(1)**

14. A sketch of a cubic function, f, with domain $-4 \leq x \leq 4$, is shown in the diagram below.

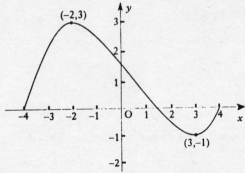

Sketch the graph of the derived function, f', for the same domain.   **(3)**

15. A medical technician obtains this print-out of a wave form generated by an oscilloscope.

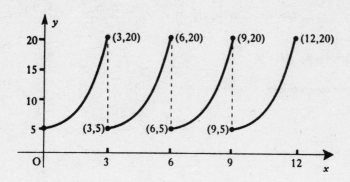

The technician knows that the equation of the first branch of the graph (for $0 \le x \le 3$) should be of the form $y = ae^{kx}$.

(a) Find the values of a and k. **(4)**

(b) Find the equation of the second branch of the curve (i.e. for $3 \le x \le 6$). **(1)**

16. Find $\displaystyle\int \sqrt{1+3x}\, dx$ and hence find the **exact** value of $\displaystyle\int_{0}^{1} \sqrt{1+3x}\, dx$ **(4)**

17. If $f(a) = 6\sin^2 a - \cos a$, express $f(a)$ in the form $p\cos^2 a + q\cos a + r$.

Hence solve, correct to three decimal places, the equation
$$6\sin^2 a - \cos a = 5 \quad \text{for} \quad 0 \le a \le \pi.$$ **(4)**

Marks

18. Explain why the equation $x^2 + y^2 + 2x + 3y + 5 = 0$ does **not** represent a circle. (2)

19. (a) Show that $(\cos x + \sin x)^2 = 1 + \sin 2x$. (1)

(b) Hence find $\int (\cos x + \sin x)^2 \, dx$. (3)

20. The point P (p,k) lies on the curve with equation $y = \log_e x$.

The point Q (q,k) lies on the curve with equation $y = \frac{1}{2} \log_e x$.

Find a relationship between p and q and hence find q when $p = 5$. (4)

21. The diagram shows the graph of the function

$$f(x) = \frac{1}{x+1}, \quad x \neq -1.$$

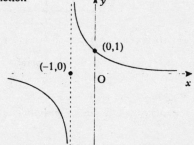

Prove that the function f is decreasing for all values of x except $x = -1$. (4)

[END OF QUESTION PAPER]

MATHEMATICS (REVISED)

Higher Grade—PAPER II

All questions should be attempted

Marks

1. The function f, whose incomplete graph is shown in the diagram, is defined by $f(x) = x^4 - 2x^3 + 2x - 1$. Find the coordinates of the stationary points and justify their nature.

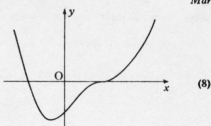

(8)

2. The concrete on the 20 feet by 28 feet rectangular facing of the entrance to an underground cavern is to be repainted.

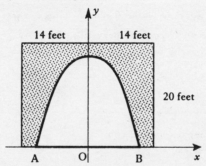

14 feet · 14 feet · 20 feet · A · O · B

Coordinate axes are chosen as shown in the diagram with a scale of 1 unit equal to 1 foot. The roof is in the form of a parabola with equation $y = 18 - \frac{1}{8}x^2$.

(a) Find the coordinates of the points A and B. (2)

(b) Calculate the total cost of repainting the facing at £3 per square foot. (4)

3. In an experiment with a ripple tank, a series of concentric circles with centre C(4,−1) is formed as shown in the diagram.

The line l with equation $y = 2x + 6$ represents a barrier placed in the tank. The largest complete circle touches the barrier at the point T.

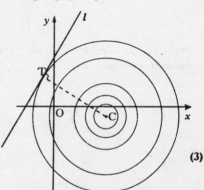

(a) Find the equation of the radius CT. (3)

(b) Find the equation of the largest complete circle. (5)

4. An array of numbers such as $\begin{pmatrix} a & b \\ c & d \end{pmatrix}$ is called a matrix.

The eigenvalues of the matrix $A = \begin{pmatrix} a & b \\ c & d \end{pmatrix}$ are defined to be the roots of the

equation $(a-x)(d-x) - bc = 0$.

EXAMPLE

In order to find the eigenvalues of the matrix $B = \begin{pmatrix} 1 & 3 \\ 4 & 2 \end{pmatrix}$

solve $\qquad (1-x)(2-x) - 4 \times 3 = 0$

solution: $\qquad 2 - 3x + x^2 - 12 = 0$

$$x^2 - 3x - 10 = 0$$
$$(x+2)(x-5) = 0$$
$$x = -2 \text{ or } x = 5$$

so the eigenvalues of B are -2 and 5

(a) Find the eigenvalues of $C = \begin{pmatrix} 3 & 4 \\ 2 & 5 \end{pmatrix}$. **(3)**

(b) Find the value of t for which the eigenvalues of the matrix $D = \begin{pmatrix} 3 & -1 \\ t & 1 \end{pmatrix}$ are

equal. **(5)**

5. A model of a crystal was made from a cube of side 3 units by slicing off the corner at P to leave a triangular face ABC.
Coordinate axes have been introduced as shown in the diagram.
The point A divides OP in the ratio 1:2.
Points B and C similarly divide RP and SP respectively in the ratio 1:2.

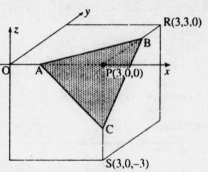

(a) Find the coordinates of A, B and C. **(3)**

(b) Calculate the area of triangle ABC. **(4)**

(c) Calculate the percentage increase or decrease in the surface area of the crystal compared with the cube. **(5)**

6. The diagram below shows the graph of $y = 2\sin 2x + 1$ for $0 \le x \le \pi$.

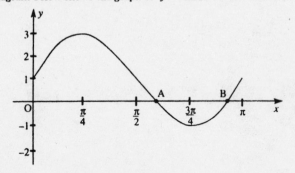

(a) Find the coordinates of A and B (as shown in the diagram) by solving an appropriate equation algebraically. **(5)**

(b) The points (0, 2) and (π, 0) are joined by a straight line l. In how many points does l intersect the given graph? **(1)**

(c) C is the point on the given graph with an x-coordinate of $\frac{\pi}{2}$. Explain whether C is above, below or on the line l. **(3)**

7. The diagram shows the plans for a proposed
new racing circuit. The designer wishes to
introduce a slip road at B for cars wishing to
exit from the circuit to go into the pits. The
designer needs to ensure that the two sections
of road touch at B in order that drivers may
drive straight on when they leave the circuit.

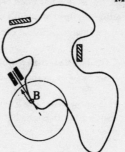

Relative to appropriate axes, the part of the circuit circled above is shown
below. This part of the circuit is represented by a curve with equation
$y = 5 - 2x^2 - x^3$ and the proposed slip road is represented by a straight line
with equation $y = -4x - 3$.

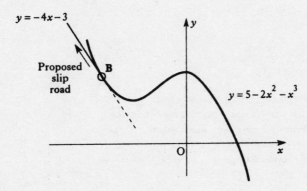

(a) Find algebraically the coordinates of B.   (7)

(b) Justify the designer's decision that this direction for the slip road does
allow drivers to go straight on.   (1)

8. Secret Agent 004 has been captured and his captors are giving him a 25 milligram dose of a truth serum every 4 hours. 15% of the truth serum present in his body is lost every hour.

(a) Calculate how many milligrams of serum remain in his body after 4 hours (that is, immediately **before** the second dose is given). **(3)**

(b) It is known that the level of serum in the body has to be continuously above 20 milligrams before the victim starts to confess. Find how many doses are needed before the captors should begin their interrogation. **(3)**

(c) Let u_n be the amount of serum (in milligrams) in his body just **after** his n^{th} dose. Show that $u_{n+1} = 0 \cdot 522 u_n + 25$. **(1)**

(d) It is also known that 55 milligrams of this serum in the body will prove fatal, and the captors wish to keep Agent 004 alive. Is there any maximum length of time for which they can continue to administer this serum and still keep him alive? **(4)**

9. A builder has obtained a large supply of 4 metre rafters. He wishes to use them to build some holiday chalets. The planning department insists that the gable end of each chalet should be in the form of an isosceles triangle surmounting two squares, as shown in the diagram.

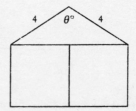

(a) If $\theta°$ is the angle shown in the diagram and A is the area (in square metres) of the gable end, show that

$$A = 8\left(2 + \sin\theta° - 2\cos\theta°\right).$$ **(5)**

(b) Express $8\sin\theta° - 16\cos\theta°$ in the form $k\sin(\theta - \alpha)°$. **(4)**

(c) Find algebraically the value of θ for which the area of the gable end is 30 square metres. **(4)**

Marks

10. When the switch in this circuit was closed, the computer printed out a graph of the current flowing (I microamps) against the time (t seconds). This graph is shown in figure 1.

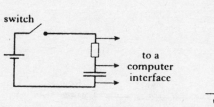

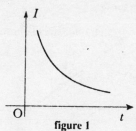

figure 1

In order to determine the equation of the graph shown in figure 1, values of $\log_e I$ were plotted against $\log_e t$ and the best fitting straight line was drawn as shown in figure 2.

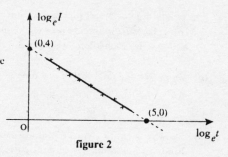

figure 2

(a) Find the equation of the line shown in figure 2 in terms of $\log_e I$ and $\log_e t$. **(3)**

(b) Hence or otherwise show that I and t satisfy a relationship of the form $I = kt^r$ stating the values of k and r. **(4)**

11. An oil production platform, 9√3 km offshore, is to be connected by a pipeline to a refinery on shore, 100 km down the coast from the platform as shown in the diagram.

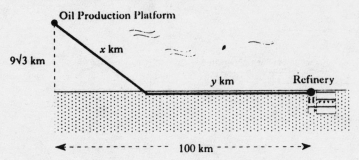

Oil Production Platform

9√3 km

x km

y km

Refinery

100 km

The length of underwater pipeline is *x* km and the length of pipeline on land is *y* km. It costs £2 million to lay each kilometre of pipeline underwater and £1 million to lay each kilometre of pipeline on land.

(a) Show that the total cost of this pipeline is £C(x) million where

$$C(x) = 2x + 100 - \left(x^2 - 243\right)^{\frac{1}{2}}.$$ (3)

(b) Show that $x = 18$ gives a minimum cost for this pipeline.
Find this minimum cost and the corresponding total length of the pipeline. (7)

[END OF QUESTION PAPER]

SCOTTISH CERTIFICATE OF EDUCATION

MATHEMATICS (REVISED)

Higher Grade—PAPER I

Tuesday, 10th May—9.30 a.m. to 11.30 a.m.

All questions should be attempted   *Marks*

1. Find $\int (3x^3 + 4x)dx$.   **(3)**

2. If $f(x) = kx^3 + 5x - 1$ and $f'(1) = 14$, find the value of k.   **(3)**

3. A is the point $(2, -1, 4)$, B is $(7, 1, 3)$ and C is $(-6, 4, 2)$. If ABCD is a parallelogram, find the coordinates of D.

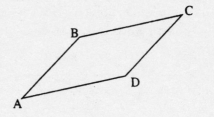

(3)

4.

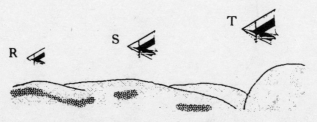

Relative to the top of a hill, three gliders have positions given by $R(-1, -8, -2)$, $S(2, -5, 4)$ and $T(3, -4, 6)$.
Prove that R, S and T are collinear.   **(3)**

5. The circle shown in the diagram has equation $(x - 1)^2 + (y - 1)^2 = 5$.

Tangents are drawn at the points $(3, 2)$ and $(2, -1)$.

Write down the coordinates of the centre of the circle and hence show that the tangents are perpendicular to each other.

(4)

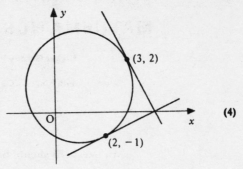

6. Find algebraically the **exact** value of $\sin\theta° + \sin(\theta + 120)° + \cos(\theta + 150)°$. **(3)**

7. If $u = \begin{pmatrix} -3 \\ 3 \\ 3 \end{pmatrix}$ and $v = \begin{pmatrix} 1 \\ 5 \\ -1 \end{pmatrix}$, write down the components of $u + v$ and $u - v$.

Hence show that $u + v$ and $u - v$ are perpendicular. **(3)**

8. The straight line $y = x$ cuts the circle $x^2 + y^2 - 6x - 2y - 24 = 0$ at A and B.

(*a*) Find the coordinates of A and B. **(3)**

(*b*) Find the equation of the circle which has AB as diameter. **(3)**

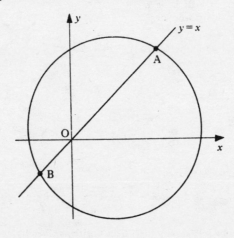

9. A sequence is defined by the recurrence relation
$$u_n = 0.9u_{n-1} + 2, \quad u_1 = 3.$$

(a) Calculate the value of u_2. **(1)**

(b) What is the smallest value of n for which $u_n > 10$? **(1)**

(c) Find the limit of this sequence as $n \to \infty$. **(2)**

10. Find the derivative, with respect to x, of $\dfrac{1}{x^3} + \cos 3x$. **(4)**

11. Show that $x^2 + 8x + 18$ can be written in the form $(x + a)^2 + b$.

Hence or otherwise find the coordinates of the turning point of the curve with equation $y = x^2 + 8x + 18$. **(3)**

12. The diagram shows the graph of the function $y = a + b\sin cx$ for $0 \le x \le \pi$.

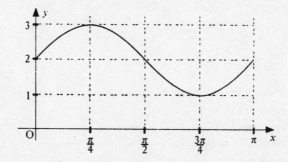

(a) Write down the values of a, b and c. **(3)**

(b) Find algebraically the values of x for which $y = 2.5$. **(3)**

45

13. If $\cos\theta = \frac{4}{5}$, $0 \leq \theta < \frac{\pi}{2}$, find the **exact** value of

    (*a*) $\sin 2\theta$ **(2)**

    (*b*) $\sin 4\theta$.   **(3)**

14. Find the gradient of the tangent to the parabola $y = 4x - x^2$ at $(0, 0)$.

Hence calculate the size of the angle between the line $y = x$ and this tangent.   **(6)**

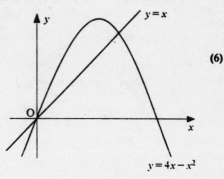

15. Solve algebraically the equation $\cos 2x° + 5\cos x° - 2 = 0$, $0 \leq x < 360$.   **(5)**

16.

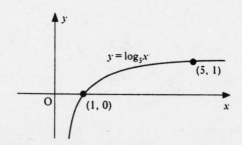

The diagram shows a sketch of part of the graph of $y = \log_5 x$.

    (*a*)   Make a copy of the graph of $y = \log_5 x$.
        On your copy, sketch the graph of $y = \log_5 x + 1$.
        Find the coordinates of the point where it crosses the x-axis.   **(3)**

    (*b*)   Make a second copy of the graph of $y = \log_5 x$.
        On your copy, sketch the graph of $y = \log_5 \frac{1}{x}$.   **(2)**

17. Differentiate $\sin^3 x$ with respect to x.

    Hence find $\int \sin^2 x \cos x \, dx$. **(4)**

18. The diagram shows a point P with coordinates $(4, 2, 6)$ and two points S and T which lie on the x-axis. If P is 7 units from S and 7 units from T, find the coordinates of S and T. **(3)**

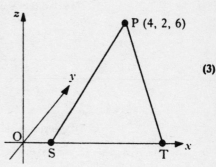

19. A function f is defined on the set of real numbers by

$$f(x) = \frac{x}{1-x}, \quad (x \neq 1).$$

    Find, in its simplest form, an expression for $f(f(x))$. **(3)**

20. The diagram shows part of the graph with equation $y = 3^x$ and the straight line with equation $y = 42$. These graphs intersect at P.

    Solve algebraically the equation $3^x = 42$, and hence write down, correct to 3 decimal places, the coordinates of P. **(4)**

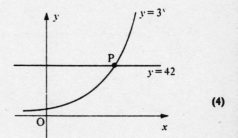

[*END OF QUESTION PAPER*]

SCOTTISH CERTIFICATE OF EDUCATION

MATHEMATICS (REVISED)

Higher Grade—PAPER II

Tuesday, 10th May 1.30 p.m. to 4.00 p.m.

All questions should be attempted

Marks

1. The graph of the curve with equation $y = 2x^3 + x^2 - 13x + a$ crosses the x-axis at the point $(2, 0)$.

   (a) Find the value of a and hence write down the coordinates of the point at which this curve crosses the y-axis. **(3)**

   (b) Find algebraically the coordinates of the other points at which the curve crosses the x-axis. **(4)**

2. ABCD is a square. A is the point with coordinates $(3, 4)$ and ODC has equation $y = \frac{1}{2}x$.

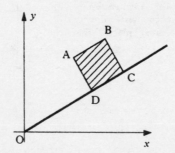

   (a) Find the equation of the line AD. **(3)**

   (b) Find the coordinates of D. **(3)**

   (c) Find the area of the square ABCD. **(2)**

3.

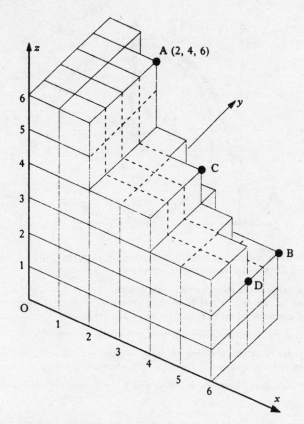

With coordinate axes as shown, the point A is (2, 4, 6).

(a) Write down the coordinates of B, C and D. **(3)**

(b) Show that C is the midpoint of AD. **(1)**

(c) By using the components of the vectors $\overrightarrow{OA}$ and $\overrightarrow{OB}$, calculate the size of angle AOB, where O is the origin. **(4)**

(d) Hence calculate the size of angle OAB. **(2)**

4.

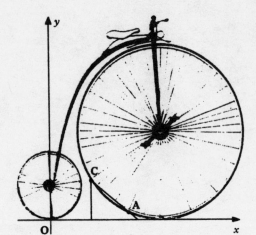

A penny-farthing bicycle on display in a museum is supported by a stand at points A and C. A and C lie on the front wheel.

With coordinate axes as shown and 1 unit = 5cm, the equation of the rear wheel (the small wheel) is $x^2 + y^2 - 6y = 0$ and the equation of the front wheel is $x^2 + y^2 - 28x - 20y + 196 = 0$.

(a) (i) Find the distance between the centres of the two wheels.

   (ii) Hence calculate the clearance, ie the smallest gap, between the front and rear wheels. Give your answer to the nearest millimetre. **(8)**

(b) B(7, 3) is half-way between A and C, and P is the centre of the front wheel.

   (i) Find the gradient of PB.

   (ii) Hence find the equation of AC and the coordinates of A and C. **(8)**

5. (a) Express $3\sin x° - \cos x°$ in the form $k\sin(x - \alpha)°$, where $k > 0$ and $0 \le \alpha \le 90$. **(4)**

   (b) Hence find algebraically the values of x between 0 and 180 for which $3\sin x° - \cos x° = \sqrt{5}$. **(4)**

   (c) Find the range of values of x between 0 and 180 for which $3\sin x° - \cos x° \le \sqrt{5}$. **(2)**

6. EXAMPLE

(i) Let $f(x) = x^3 + 5x - 1$.

Since $f(0) = -1$ and $f(1) = 5$,
the equation $f(x) = 0$ has a root in the interval $0 < x < 1$ because $f(0) < 0$ and $f(1) > 0$.

(ii) To find this root, the equation $x^3 + 5x - 1 = 0$ can be rearranged as follows:

$$x^3 + 5x - 1 = 0$$
$$x^3 + 5x = 1$$
$$x(x^2 + 5) = 1$$
$$x = \frac{1}{x^2 + 5}$$

We can write this result as a recurrence relation

$$x_{n+1} = \frac{1}{x_n^2 + 5}$$

and use it to find this root. In this example we will work to 3 decimal places and can therefore give the final answer to 2 decimal places.

(iii) For our first estimate, x_1, we use the mid-point of the interval $0 < x < 1$ [from part (i)].

$$x_1 = 0.5, \qquad x_2 = \frac{1}{0.5^2 + 5} \qquad = 0.190$$

$$x_2 = 0.190 \qquad x_3 = \frac{1}{0.190^2 + 5} \qquad = 0.199$$

$$x_3 = 0.199 \qquad x_4 = \frac{1}{0.199^2 + 5} \qquad = 0.198$$

$$x_4 = 0.198 \qquad x_5 = \frac{1}{0.198^2 + 5} \qquad = 0.198$$

Hence, correct to 2 decimal places, the root is $x = 0.20$.

(a) Show that the equation $2x^3 + 3x - 1 = 0$ has a root in the interval $0 < x < 0.5$.

(2)

(b) By using the technique described above, find this root correct to 2 decimal places.

(6)

7. A yacht club is designing its new flag.

The flag is to consist of a red triangle on a yellow rectangular background.

In the yellow rectangle ABCD, AB measures 8 units and AD is 6 units. E and F lie on BC and CD, x units from B and C as shown in the diagram.

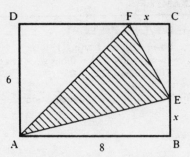

(a) Show that the area, H square units, of the red triangle AEF is given by $H(x) = 24 - 4x + \frac{1}{2}x^2$. **(4)**

(b) Hence find the greatest and least possible values of the area of triangle AEF. **(8)**

8. (a) $f(x) = 4x^2 - 3x + 5$.

Show that $f(x + 1)$ simplifies to $4x^2 + 5x + 6$ and find a similar expression for $f(x - 1)$.

Hence show that $\dfrac{f(x + 1) - f(x - 1)}{2}$ simplifies to $8x - 3$. **(5)**

(b) $g(x) = 2x^2 + 7x - 8$.

Find a similar expression for $\dfrac{g(x + 1) - g(x - 1)}{2}$. **(4)**

(c) By examining your answers for (a) and (b), **write down** the simplified expression for $\dfrac{h(x + 1) - h(x - 1)}{2}$, where $h(x) = 3x^2 - 5x - 1$. **(2)**

9. (*a*) The point A(2, 2) lies on the parabola $y = x^2 + px + q$.

Find a relationship between p and q.   **(1)**

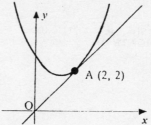

(*b*) The tangent to the parabola at A is the line $y = x$. Find the value of p. Hence find the equation of the parabola.   **(6)**

(*c*) Using your answers for p and q, find the value of the discriminant of $x^2 + px + q = 0$. What feature of the above sketch is confirmed by this value?   **(2)**

10. The cargo space of a small bulk carrier is 60m long.

The shaded part of the diagram below represents the uniform cross-section of this space. It is shaped like the parabola with equation $y = \frac{1}{4}x^2$, $-6 \le x \le 6$, between the lines $y = 1$ and $y = 9$.

Find the area of this cross-section and hence find the volume of cargo that this ship can carry.   **(9)**

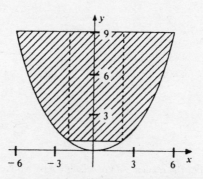

[END OF QUESTION PAPER]

SCOTTISH
CERTIFICATE OF
EDUCATION
1995

TUESDAY, 9 MAY
9.30 AM – 11.30 AM

MATHEMATICS (REVISED)
HIGHER GRADE
Paper I

All questions should be attempted

Marks

1. Calculate the length of the vector $2i - 3j + \sqrt{3}k$. **(2)**

2. (a) Show that $(x - 3)$ is a factor of $f(x)$, where $f(x) = 2x^3 + 3x^2 - 23x - 12$. **(2)**

   (b) Hence express $f(x)$ in its fully factorised form. **(2)**

3. Find $\int (6x^2 - x + \cos x)\, dx$. **(4)**

4. Find the value of k for which the vectors $\begin{pmatrix} 1 \\ 2 \\ -1 \end{pmatrix}$ and $\begin{pmatrix} -4 \\ 3 \\ k - 1 \end{pmatrix}$ are perpendicular. **(3)**

5. Find the equation of the median AD of triangle ABC where the coordinates of A, B and C are $(-2, 3)$, $(-3, -4)$ and $(5, 2)$ respectively. **(3)**

6. A Royal Navy submarine, exercising in the Firth of Clyde, is stationary on the seabed below a point S on the surface. S is the point (5, 4) as shown in the diagram.

   A radar operator observes the frigate "Achilles" sailing in a straight line, passing through the points A_1 (−4, −1) and A_2 (−1, 1).

   Similarly, the frigate "Belligerent" is observed sailing in a straight line, passing through the points B_1 (−7,−11) and B_2 (1, −1).

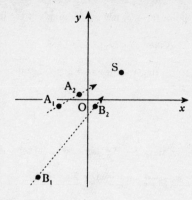

   If both frigates continue to sail in straight lines, will either or both frigates pass directly over the submarine?

   (5)

7. Find $\dfrac{dy}{dx}$ where $y = \dfrac{4}{x^2} + x\sqrt{x}$.

   (4)

8. Find the exact solutions of the equation

   $$4\sin^2 x = 1, \quad 0 \le x < 2\pi.$$

   (4)

9. Find the equation of the circle which has P(−2, −1) and Q(4, 5) as the end points of a diameter.

   (3)

10. The point $P(-2, b)$ lies on the graph of the function $f(x) = 3x^3 - x^2 - 7x + 4$.

   (*a*) Find the value of b.   **(1)**

   (*b*) Prove that this function is increasing at P.   **(3)**

11. The functions f and g, defined on suitable domains, are given by

$$f(x) = \frac{1}{x^2 - 4} \text{ and } g(x) = 2x + 1.$$

   (*a*) Find an expression for $h(x)$ where $h(x) = g(f(x))$. Give your answer as a single fraction.   **(3)**

   (*b*) State a suitable domain for h.   **(1)**

12. Given that $\tan \alpha = \frac{\sqrt{11}}{3}$, $0 < \alpha < \frac{\pi}{2}$, find the exact value of $\sin 2\alpha$.   **(3)**

13. Solve the simultaneous equations

$$k \sin x° = 5$$

$$k \cos x° = 2 \text{ where } k > 0 \text{ and } 0 \le x \le 360.$$ **(4)**

14. The straight line shown in the diagram has equation $y = f(x)$.

Determine $f'(x)$.

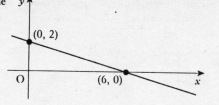

(2)

15. Solve algebraically the equation

$$\cos 2x° + \cos x° = 0, \quad 0 \le x < 360.$$

(5)

16. In the square-based pyramid, all the eight edges are of length 3 units.

$\overrightarrow{AV} = \boldsymbol{p}, \quad \overrightarrow{AD} = \boldsymbol{q}, \quad \overrightarrow{AB} = \boldsymbol{r}.$

Evaluate $\boldsymbol{p}.(\boldsymbol{q} + \boldsymbol{r})$.

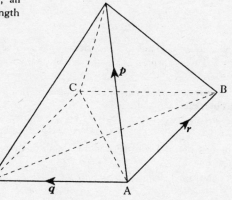

(4)

17. Part of the graph of $y = f(x)$ is shown in the diagram. This graph has stationary points at $x = 0$, $x = a$ and $x = b$.

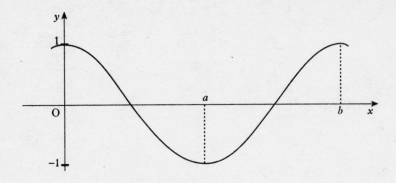

   (a) Sketch the graph of $y = f'(x)$ for $0 \leq x \leq b$.   **(3)**

   (b) If $a = \pi$ and $b = 2\pi$, write down a possible expression for $f'(x)$.   **(1)**

18. The amount A grams of a radioactive substance after a time t minutes is given by $A = A_0 e^{-kt}$ where A_0 is the initial amount of the substance and k is a constant.

In 3 minutes, 10 grams of the substance Bismuth are reduced to 9 grams through radioactive decay.

   (a) Find the value of k.   **(3)**

The half-life of a substance is the length of time in which half the substance decays.

   (b) Find the half-life of Bismuth.   **(2)**

19. The diagram shows a sketch of the graph of $y = f(x)$, where $f(x) = a \log_2(x - b)$.

Find the values of a and b.

(3)

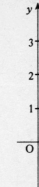

20. The roots of the equation $(x - 1)(x + k) = -4$ are equal.

Find the values of k.

(5)

21. A ball is thrown vertically upwards. The height h metres of the ball t seconds after it is thrown, is given by the formula $h = 20t - 5t^2$.

(a) Find the speed of the ball when it is thrown (i.e. the rate of change of height with respect to time of the ball when it is thrown).

(3)

(b) Find the speed of the ball after 2 seconds.

Explain your answer in terms of the movement of the ball.

(2)

[END OF QUESTION PAPER]

SCOTTISH
CERTIFICATE OF
EDUCATION
1995

TUESDAY, 9 MAY
1.30 PM – 4.00 PM

MATHEMATICS (REVISED)
HIGHER GRADE
Paper II

All questions should be attempted

Marks

1. A triangle ABC has vertices A(4, 8), B(1, 2) and C(7, 2).

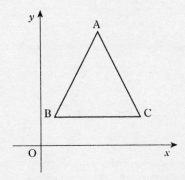

(a) Show that the triangle is isosceles. **(2)**

(b) (i) The altitudes AD and BE intersect at H, where D and E lie on BC and CA respectively. Find the coordinates of H. **(7)**

(ii) Hence show that H lies one quarter of the way up DA. **(1)**

2. The diagram shows a sketch of part of the graph of $y = x^3 - 2x^2 + x$.

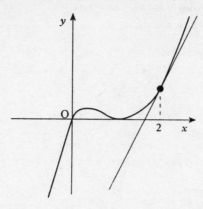

(a) Show that the equation of the tangent to the curve at $x = 2$ is $y = 5x - 8$. **(4)**

(b) Find algebraically the coordinates of the point where this tangent meets the curve again. **(5)**

3. Trees are sprayed weekly with the pesticide, "Killpest", whose manufacturers claim it will destroy 65% of all pests. Between the weekly sprayings, it is estimated that 500 new pests invade the trees.

A new pesticide, "Pestkill", comes onto the market. The manufacturers claim that it will destroy 85% of existing pests but it is estimated that 650 new pests per week will invade the trees.

Which pesticide will be more effective in the long term? **(7)**

4. (*a*) (i) Diagram 1 shows part of the graph of the function *f* defined by $f(x) = b \sin ax°$, where *a* and *b* are constants.

Write down the values of *a* and *b*.

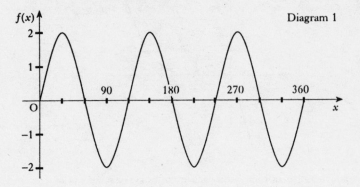

Diagram 1

(ii) Diagram 2 shows part of the graph of the function *g* defined by $g(x) = d \cos cx°$, where *c* and *d* are constants.

Write down the values of *c* and *d*. **(4)**

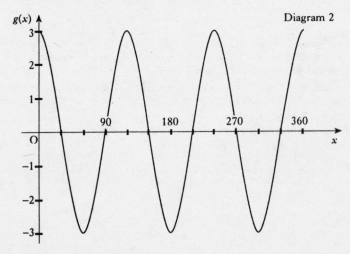

Diagram 2

(*b*) The function *h* is defined by $h(x) = f(x) + g(x)$.

Show that $h(x)$ can be expressed in terms of a single trigonometric function of the form $q \sin(px + r)°$ and find the values of *p*, *q* and *r*. **(5)**

5. The diagram shows the rhombohedral crystal lattice of calcium carbonate.

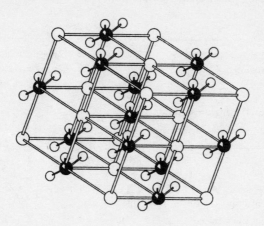

The three oxygen atoms P, Q and R around the carbon atom A have coordinates as shown.

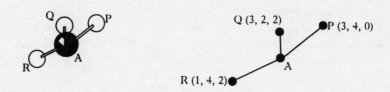

(a) Calculate the size of angle PQR. **(4)**

(b) M is the midpoint of QR and T is the point which divides PM in the ratio 2:1.
  (i) Find the coordinates of T.
  (ii) Show that P, Q and R are equidistant from T. **(6)**

(c) The coordinates of A are (2, 3, 1).
  (i) Show that P, Q and R are also equidistant from A.
  (ii) Explain why T, and not A, is the centre of the circle through P, Q and R. **(2)**

6. A system of 3 equations in 3 unknowns can be solved by a method known as Gaussian Elimination as shown below.

Example

Solve the system of equations
by Gaussian Elimination.

$$x + 2y - 3z = 11$$
$$2x + 2y - z = 11$$
$$3x - 2y + 4z = -4$$

A   Write out the coefficients in an array:

- Row 1
- Row 2
- Row 3

1	2	-3	11
2	2	-1	11
3	-2	4	-4

B   Keep Row 1 the same. Make Row 2 and Row 3 each begin with a zero by subtracting multiples of Row 1 from them.

- Row 1 is kept the same
- Row 2 becomes "Row 2 – 2 × Row 1"
- Row 3 becomes "Row 3 – 3 × Row 1"

1	2	-3	11
0	-2	5	-11
0	-8	13	-37

C   Keep Row 1 and Row 2 the same. Make Row 3 begin with two zeros, by subtracting a multiple of Row 2 from it.

- Row 1 is kept the same ....(1)
- Row 2 is kept the same ....(2)
- Row 3 becomes "Row 3 – 4 × Row 2" ....(3)

1	2	-3	11
0	-2	5	-11
0	0	-7	7

D

- Line (3) gives $\qquad -7z = 7, \qquad z = -1$

- Line (2) gives
$$-2y + 5z = -11$$
$$-2y + (-5) = -11, \qquad y = 3$$

- Line (1) gives
$$x + 2y - 3z = 11$$
$$x + 6 + 3 = 11, \qquad x = 2$$

So the solution is $x = 2$, $y = 3$, $z = -1$

Solve the following system of equations by
Gaussian Elimination **as shown above**.

$$x - 2y + z = 6$$
$$3x + y - z = 7$$
$$4x - y + 2z = 15$$

(7)

Marks

7. The parabola $y = ax^2 + bx + c$ crosses the y-axis at $(0, 3)$ and has two tangents drawn, as shown in the diagram.

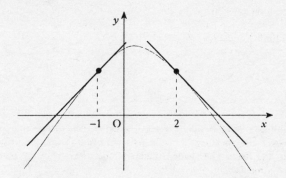

The tangent at $x = -1$ makes an angle of $45°$ with the positive direction of the x-axis and the tangent at $x = 2$ makes an angle of $135°$ with the positive direction of the x-axis.

Find the values of a, b and c. **(8)**

8. When newspapers were printed by lithograph, the newsprint had to run over three rollers, illustrated in the diagram by three circles. The centres A, B and C of the three circles are collinear.

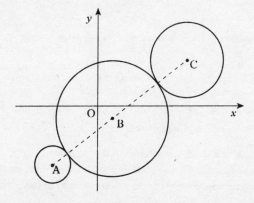

The equations of the circumferences of the outer circles are
$(x + 12)^2 + (y + 15)^2 = 25$ and $(x - 24)^2 + (y - 12)^2 = 100$.

Find the equation of the central circle. **(8)**

Marks

9. (a) By writing sin $3x$ as sin $(2x + x)$, show that

sin $3x = 3$ sin $x - 4$ sin^3 x. **(4)**

(b) Hence find $\int \sin^3 x\, dx$. **(4)**

10. When building a road beside a vertical rockface, engineers often use wire mesh to cover the rockface. This helps to prevent rocks and debris from falling onto the road. The shaded region of the diagram below represents a part of such a rockface.

This shaded region is bounded by a parabola and a straight line.

The equation of the parabola is $y = 4 + \frac{5}{3}x - \frac{1}{6}x^2$ and the equation of the line is $y = 4 - \frac{1}{3}x$.

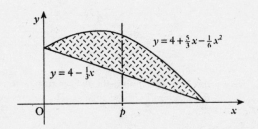

(a) Find algebraically the area of wire mesh required for this part of the rockface. **(5)**

(b) To help secure the wire mesh, weights are attached to the mesh along the line $x = p$ so that the area of mesh is bisected.

By using your answer to part (a), or otherwise, show that p satisfies the equation $p^3 - 18p^2 + 432 = 0$. **(3)**

(c) (i) Verify that $p = 6$ is a solution of this equation.

(ii) Find algebraically the other two solutions of this equation.

(iii) Explain why $p = 6$ is the only valid solution to this problem. **(5)**

11. Linktown Church is considering designs for a logo for their parish magazine. The "C" is part of a circle and the centre of the circle is the mid-point of the vertical arm of the "L". Since the "L" is clearly smaller than the "C", the designer wishes to ensure that the total length of the arms of the "L" is as long as possible.

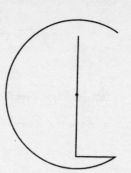

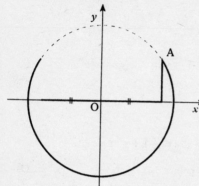

The designer decides to call the point where the "L" and "C" meet A and chooses to draw coordinate axes so that A is in the first quadrant. With axes as shown, the equation of the circle is $x^2 + y^2 = 20$.

(a) If A has coordinates (x, y), show that the total length T of the arms of the "L" is given by $T = 2x + \sqrt{20 - x^2}$.

(1)

(b) Show that for a stationary value of T, x satisfies the equation

$$x = 2\sqrt{20 - x^2}.$$

(5)

(c) By squaring both sides, solve this equation.

Hence find the greatest length of the arms of the "L".

(3)

[END OF QUESTION PAPER]

67

SCOTTISH
CERTIFICATE OF
EDUCATION
1996

WEDNESDAY, 8 MAY
9.30 AM – 11.30 AM

MATHEMATICS
HIGHER GRADE
Paper I

All questions should be attempted

Marks

1. Find the equation of the perpendicular bisector of the line joining A(2, −1) and B(8, 3). **(4)**

2. For what value of a does the equation $ax^2 + 20x + 40 = 0$ have equal roots? **(2)**

3. The diagram shows an incomplete graph of $y = 3 \sin\left(x - \dfrac{\pi}{3}\right)$, for $0 \leq x \leq 2\pi$.

   Find the coordinates of the maximum stationary point. **(3)**

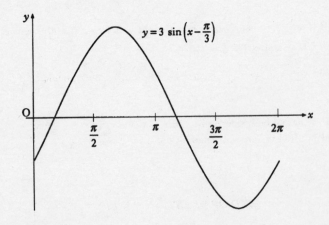

4. Find the equation of the tangent at the point (3, 4) on the circle

   $$x^2 + y^2 + 2x - 4y - 15 = 0.$$ **(4)**

5. Evaluate $\displaystyle\int_{-3}^{0} (2x+3)^2 \, dx$. **(4)**

Marks

6. A is the point $(2, -5, 6)$, B is $(6, -3, 4)$ and C is $(12, 0, 1)$. Show that A, B and C are collinear and determine the ratio in which B divides AC. **(4)**

7. Express $x^4 - x$ in its fully factorised form. **(4)**

8. Part of the graph of $y = f(x)$ is shown in the diagram. On separate diagrams, sketch the graphs of

   (i) $y = f(x - 1)$

   (ii) $y = -f(x) - 2$

   indicating on each graph the images of A, B, C and D. **(5)**

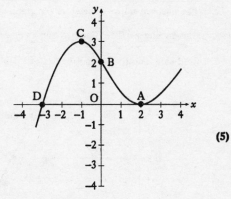

9. Find $f'(4)$ where $f(x) = \dfrac{x-1}{\sqrt{x}}$. **(5)**

10. Solve algebraically the equation

$$\sin 2x° + \sin x° = 0, \qquad 0 \le x < 360.$$ **(5)**

Marks

11. A sequence is defined by the recurrence relation $u_{n+1} = 0 \cdot 3 u_n + 5$ with first term u_1.

    (a) Explain why this sequence has a limit as n tends to infinity. **(1)**

    (b) Find the **exact** value of this limit. **(2)**

12. The diagram shows a sketch of the graph of $y = \sin\left(2x - \dfrac{\pi}{6}\right)$, $0 \leq x \leq \pi$, and the straight line $y = 0 \cdot 5$. These graphs intersect at P and Q.

    Find algebraically the coordinates of P and Q. **(4)**

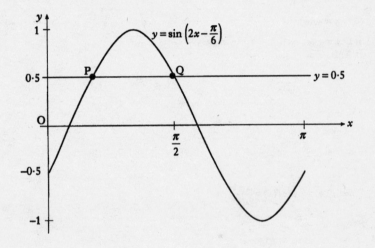

13. Find $\dfrac{dy}{dx}$, given that $y = \sqrt{1 + \cos x}$. **(3)**

14. Three lines have equations $2x + 3y - 4 = 0$, $3x - y - 17 = 0$ and $x - 3y - 10 = 0$.

    Determine whether or not these lines are concurrent. **(4)**

Marks

15. The diagram shows two right-angled triangles ABD and BCD with AB = 7 cm, BC = 4 cm and CD = 3 cm. Angle DBC = $x°$ and angle ABD = $y°$.

Show that the exact value of

$\cos(x+y)°$ is $\dfrac{20-6\sqrt6}{35}$.

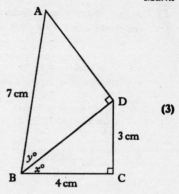

(3)

16. Find algebraically the values of x for which the function $f(x) = 2x^3 - 3x^2 - 36x$ is increasing.

(4)

17. Express $f(x) = (2x - 1)(2x + 5)$ in the form $a(x + b)^2 + c$.

(3)

18. The framework of a child's swing has dimensions as shown in the diagram on the right.

Find the exact value of $\sin x°$.

(5)

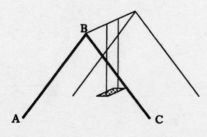

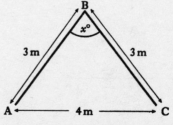

19. A mug of tea cools according to the law $T_t = T_0 e^{-kt}$, where T_0 is the initial temperature and T_t is the temperature after t minutes. All temperatures are in °C.

    (a) A particular mug of tea cooled from boiling point (100 °C) to 75 °C in a quarter of an hour. Calculate the value of k.

    (b) By how many degrees will the temperature of this tea fall in the next quarter of an hour?　　　　　　　　　　　　　　　　　　　　(5)

20. The line $y = -1$ is a tangent to a circle which passes through (0, 0) and (6, 0). Find the equation of this circle.　　　　　　　　　　　　　　　　(6)

[END OF QUESTION PAPER]

SCOTTISH
CERTIFICATE OF
EDUCATION
1996

WEDNESDAY, 8 MAY
1.30 PM – 4.00 PM

MATHEMATICS
HIGHER GRADE
Paper II

All questions should be attempted

Marks

1. A curve has equation $y = x^4 - 4x^3 + 3$.

   (a) Find algebraically the coordinates of the stationary points.   **(6)**

   (b) Determine the nature of the stationary points.   **(2)**

2. A triangle ABC has vertices A(–3, –3), B(–1, 1) and C(7, –3).

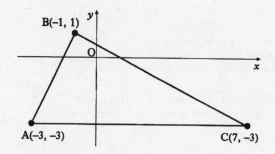

   (a) Show that the triangle ABC is right-angled at B.   **(3)**

   (b) The medians AD and BE intersect at M.

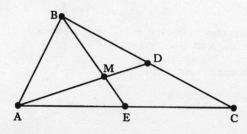

      (i) Find the equations of AD and BE.   **(5)**
      (ii) Hence find the coordinates of M.   **(3)**

Marks

3. The first four levels of a stepped pyramid with a square base are shown in the diagram.

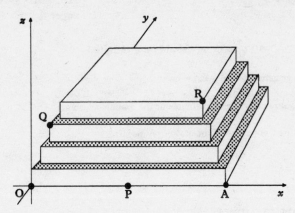

Each level is a square-based cuboid with a height of 3 m. The shaded parts indicate the steps which have a "width" of 1 m.

The height and "width" of a step at a corner are shown in this enlargement.

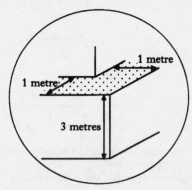

With coordinate axes as shown and 1 unit representing 1 metre, the coordinates of P and A are (12, 0, 0) and (24, 0, 0).

(a) Find the coordinates of Q and R. (2)

(b) Find the size of angle QPR. (7)

1996

Marks

4. (a) $f(x) = 2x + 1$, $g(x) = x^2 + k$, where k is a constant.
    (i) Find $g(f(x))$. (2)
    (ii) Find $f(g(x))$. (2)

    (b) (i) Show that the equation $g(f(x)) - f(g(x)) = 0$ simplifies to $2x^2 + 4x - k = 0$. (2)
    (ii) Determine the nature of the roots of this equation when $k = 6$. (2)
    (iii) Find the value of k for which $2x^2 + 4x - k = 0$ has equal roots. (3)

5. An artist has designed a "bow" shape which he finds can be modelled by the shaded area below. Calculate the area of this shape. (6)

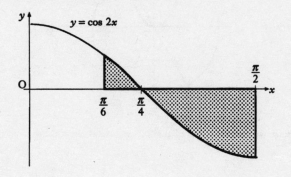

Marks

6. Diagram 1 shows:

- the point A(1, 2),
- the straight line *l* passing through the origin O and the point A,
- the parabola *p* with a minimum turning point at O and passing through A,
- and the circle *c*, centre O, passing through A.

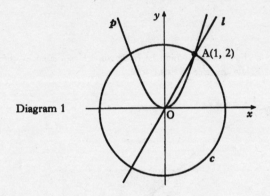

Diagram 1

(*a*) Write down the equations of the line, the parabola and the circle. **(3)**

Marks

6. **(continued)**

The following transformations are carried out:

> • the line is given a translation of 4 units down (ie −4 units in the direction of the y-axis),
>
> Diagram 2 shows the line *l′*, the image of line *l*, after this translation.
>
> • the parabola is reflected in the x-axis,
>
> • the circle is given a translation of 2 units to the right (ie +2 units in the direction of the x-axis).

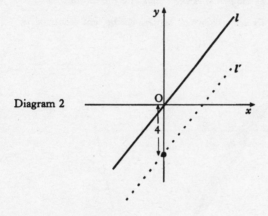

Diagram 2

(b) Write down the equations of *l′*, *p′*(the image of the parabola *p*) and *c′* (the image of the circle *c*). **(4)**

(c) (i) Show that the line *l′* passes through the centre of the circle *c′*. **(1)**

    (ii) Find the coordinates of the points where the line *l′* intersects the parabola *p′*. **(3)**

Marks

7. $f(x) = 2 \cos x° + 3 \sin x°$.

   (a) Express $f(x)$ in the form $k\cos(x-\alpha)°$ where $k > 0$ and $0 \leq \alpha < 360$.   **(4)**

   (b) Hence solve algebraically $f(x) = 0 \cdot 5$ for $0 \leq x < 360$.   **(3)**

8. In the diagram below, a winding river has been modelled by the curve $y = x^3 - x^2 - 6x - 2$ and a road has been modelled by the straight line AB. The road is a tangent to the river at the point A(1, −8).

   (a) Find the equation of the tangent at A and hence find the coordinates of B.   **(8)**

   (b) Find the area of the shaded part which represents the land bounded by the river and the road.   **(3)**

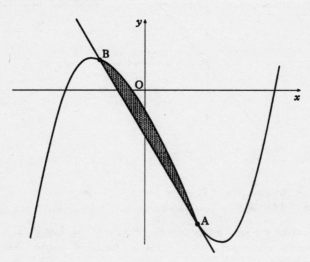

Marks

9. Six spherical sponges were dipped in water and weighed to see how much water each could absorb. The diameter (x millimetres) and the gain in weight (y grams) were measured and recorded for each sponge. It is thought that x and y are connected by a relationship of the form $y = ax^b$.

By taking logarithms of the values of x and y, the table below was constructed.

$X (= \log_e x)$	2·10	2·31	2·40	2·65	2·90	3·10
$Y (= \log_e y)$	7·00	7·60	7·92	8·70	9·38	10·00

A graph was drawn and is shown below.

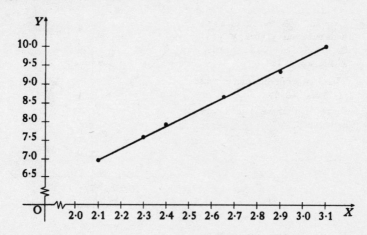

(a) Find the equation of the line in the form $Y = mX + c$. **(3)**

(b) Hence find the values of the constants a and b in the relationship $y = ax^b$. **(4)**

Marks

10. Two curves, $y = f(x)$ and $y = g(x)$, are called orthogonal if, at each point of intersection, their tangents are at right angles to each other.

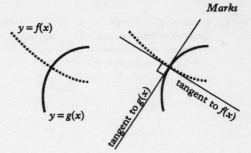

(a) Diagram 1 shows the parabola with equation $y = 6 + \frac{1}{9}x^2$ and the circle M with equation $x^2 + (y-5)^2 = 13$.

These two curves intersect at (3, 7) and (−3, 7).

Prove that these curves are orthogonal.

(6)

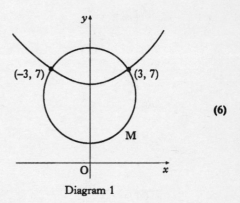

Diagram 1

(b) Diagram 2 shows the circle M, from (a) above, which is orthogonal to the circle N. The circles intersect at (3, 7) and (−3, 7).

  (i) Write down the equation of the tangent to circle M at the point (−3, 7).

(1)

  (ii) Hence find the equation of circle N.

(3)

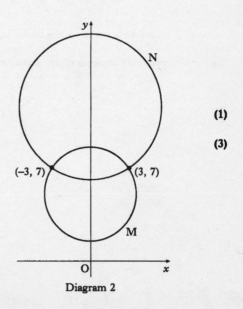

Diagram 2

11. A window in the shape of a rectangle surmounted by a semicircle is being designed to let in the maximum amount of light.

The glass to be used for the semicircular part is stained glass which lets in one unit of light per square metre; the rectangular part uses clear glass which lets in 2 units of light per square metre.

The rectangle measures $2x$ metres by h metres.

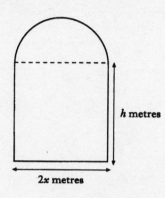

h metres

$2x$ metres

(*a*) (i) If the perimeter of the whole window is 10 metres, express h in terms of x. **(2)**

  (ii) Hence show that the amount of light, L, let in by the window is given by $L = 20x - 4x^2 - \frac{3}{2}\pi x^2$. **(2)**

(*b*) Find the values of x and h that must be used to allow this design to let in the maximum amount of light. **(5)**

[END OF QUESTION PAPER]

1. $p = -7$

2. $y = 12x + 6$

3. $2y = 3x - 1$

4. $\overrightarrow{PQ} = 2\overrightarrow{QR}; 2:1$

5. (a) $\overrightarrow{BC} = \begin{array}{c} 4 \\ 2 \\ -3 \end{array}$ (b) $\sqrt{29} \ (= 5\cdot4)$

6. 11 units^2

7. $x^2 + y^2 - 10x - 26y + 185 = 0$

8. $y = 3 + x - x^2$

9. $\dfrac{1}{\sqrt{5}} : \dfrac{3}{5}$

10. (a) $y = 2 \sin 4x°$
    (b) $A(57\cdot2°, -1\cdot5); B(77\cdot9°, -1\cdot5)$

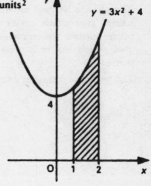

11.

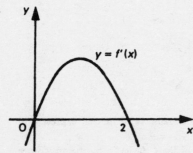

12. $a = 2: b = 3$

13. (a) $(-1, 2); (1, 2)$ (b) $\dfrac{16}{3} \text{ units}^2$

14. $y = 10x^2$

15. $\dfrac{\pi}{3}, \dfrac{2\pi}{3}$

16. $x < -1 \text{ or } x > 5$

17.

18. $b^2 - 4ac = (k + 2)^2$

19. $f'(x) = \dfrac{4}{3x^3} - \sin 2x$

20. $-\dfrac{63}{16} \ (= -3\cdot94)$

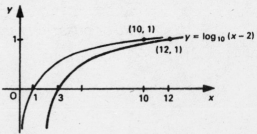

1. *(a)* $(-2, 0)$, $(1, 0)$; $(0, 2)$ *(b)* Maximum T.P. $(-1, 4)$; Minimum T.P. $(1, 0)$

   (c)

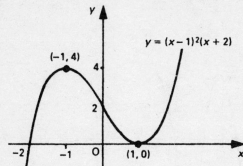

2. *(a)* $2x - y = 4$; $x + 2y = 12$; $(4, 4)$ *(b)* $(2, 5)$

3. *(a)* 73 *(b)* 33 4. *(a)* $(13, 13, 8)$ *(b)* $38 \cdot 6°$

5. *(a)* Proof *(b)* $\sqrt{7} \cos(x + 49 \cdot 1)°$; $k = \sqrt{7}$, $\alpha = 49 \cdot 1°$ *(c)* $18 \cdot 7$, $243 \cdot 1$

6. *(a)* Proof *(b)* $f(g(x)) = (x + 1)(x - 2)(x - 4)$ *(c)* $-1, 2$ or 4

7. *(a)* $\int_k^1 (\sqrt{x} - x^2) dx$ and $\int_0^k (\sqrt{x} - x^2) dx$ *(b)* Proof *(c)* $\left(1 - \dfrac{1}{\sqrt{2}}\right)^{2/3} (= 0 \cdot 44)$

8. *(a)* Proof *(b)* Tangent at $(1, 3)$

9. *(a)* $f(0) < 0$ and $f(1) > 0$ *(b)* $x = 0 \cdot 8$

10. *(a)*

$w(t)$

$w(t) = -\sin \frac{\pi}{6} t$

O 6 12 18 24 30 36 t

$w(t) = \sin \frac{\pi}{6} t$

(b)

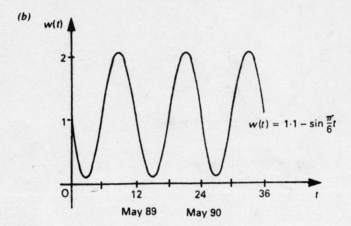

$$w(t) = 1.1 - \sin\frac{\pi}{6}t$$

(c) 28th June 1990 ($t \doteqdot 25 \cdot 94$)

PAPER I

1. $3x + 2y + 1 = 0$
2. $2b + a$
3. Proof $(\underline{a} \cdot \underline{b} = 0)$
4. (a) $k = 3$ (b) $(1, 4 \cdot 95)$
5. $y = 6x - 1$
6. $p = 8; q = 23$
7. (a) Proof $(\overrightarrow{LM} = k\overrightarrow{MN})$ (b) $4 : 1$
8. $x + 4y = 7$
9. (a)

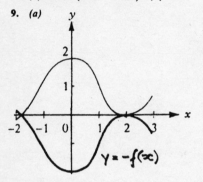

(b)

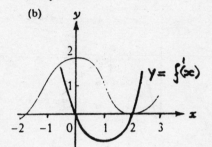

10. $f(x) = x^4 - x - 15$
11. 54%
12. $\sin 2A = \dfrac{3\sqrt{7}}{8}$
13. $f'(4) = -30$
14. (a) $-2 \leqslant x \leqslant 2$ (b) $-4 < x < -2$
15. (a) $8 - (x + 1)^2; a = 8; b = 1$ (b) 8 (when $x = -1$)
16. (a) $\dfrac{5}{3}$ (b)

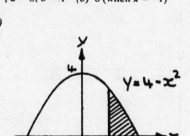

17. 8
18. Proof
19. (a) $p = 1; r = 46$ (b) $h(x) = 4x^2 + 20x + 22$
20. At A, $x = 210$

1. *(a)* Max T.P. $(0, 0)$; Min T.P. $(2, -8)$ *(b)* $-8 < k < 0$
2. *(a)* $y = 3x - 11$ *(b)* Centre $(3, -2)$; Equation is $(x - 3)^2 + (y + 2)^2 = 25$
3. *(a)* Proof (Sine Rule)   *(b)* (i) $x° = 60°$ (ii) $\sin x° = \sin 2x° \Rightarrow x = 60$
4. *(a)*   *(b)* $x = 1\cdot76$ to 2 d.p.

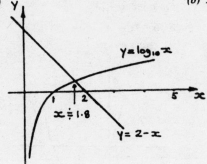

5. *(a)* (i) $\vec{VF} = \begin{pmatrix} 1 \\ 1 \\ -10 \end{pmatrix}$; $\vec{VE} = \begin{pmatrix} 1 \\ -1 \\ -10 \end{pmatrix}$ (ii) $11\cdot4°$ *(b)* $10\cdot1$ cm^2
6. $3x^2 \cos x - x^3 \sin x$ 7. *(a)* $k = 0\cdot067$ *(b)* Yes $(P_4 > 30)$
8. *(a)* $d = 2 \cos (20t - 60)°$ *(b)*

    (c) $t = 0\cdot9$ or $5\cdot1$

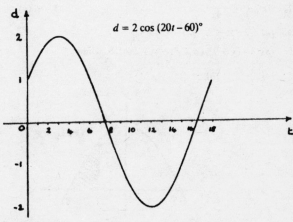

9. *(a)* $23\cdot2$ *(b)* Yes the doctor should prescribe the treatment. ($U_\infty < 93\cdot4$)
10. *(a)* Proof $(y = k(x - 20)(x + 20)$ at $(0, 40))$ *(b)* Proof (touches $\Rightarrow$ tangent)
11. *(a)* $2y = x + 1$ *(b)* $\frac{3}{4}$ units^2 *(c)* $\frac{2}{3}$ units^2
    (d) The objection is correct. The reduction is $11\cdot1\%$.

PAPER I

1. $y = 3x - 4$

2. *(a)* CE: $y = -3x + 5$; BD: $y = x + 1$ *(b)* $J = (1, 2)$

3. $k = 3$ 4. $y = x^3 + x + 4$

5. $x = 235 \cdot 3$ 6. $f\big(g(x)\big) = x$

7. *(a)* $\sin x° - 3 \cos x° = \sqrt{10} \sin (x - 71 \cdot 6)°$; $k = \sqrt{10}$, $\alpha = 71 \cdot 6$
   (b) Maximum value $= 5 + \sqrt{10}$ where $x = 161 \cdot 6$

8. $\dfrac{64}{3}$ 9. 45 m

10. *(a)* graph of $y = g(x)$

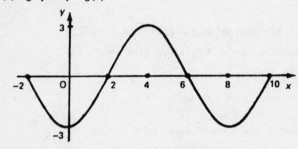

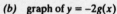

(b) graph of $y = -2g(x)$

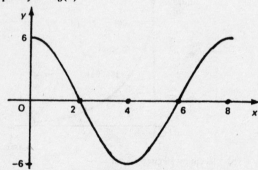

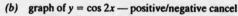

11. $3x^{1/2} + 2 \sin x \cos x$ (or $3\sqrt{x} + \sin 2x$)

12. 3   13. $3 \cdot 43$

14. *(a)* 0   *(b)* graph of $y = \cos 2x$ — positive/negative cancel

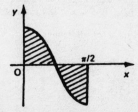

15. $C = (5, 0, -5)$

16. $\frac{25}{2} \sqrt{3}\ un^2$ (or $21 \cdot 6\ un^2$)

17. $k = 5$

18. 7

19.

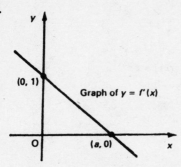

Graph of $y = f'(x)$

$(0, 1)$

$(a, 0)$

REVISED HIGHER — PAPER II

1. (a) $f(x) = (x + 4)(x + 1)(x - 4)$ (b) $(-4, 0), (-1, 0), (4, 0)$ and $(0, -16)$

    (c) Maximum at $\left(\frac{-8}{3}, \frac{400}{27}\right)$. Minimum at $(2, -36)$

2. (a) $|\overrightarrow{BR}| = 2\sqrt{29}$ km ($\doteq 16 \cdot 6$ km) (b) 429 km/hr

    (c) Proof ($\overrightarrow{BR}.\overrightarrow{TC} = 0$) (d) $36 \cdot 7°$

3. (a) Show fixed level > 5 mg/l (b) show fixed level < 5 mg/l $\Rightarrow$ yes

4. (a) $409 \cdot 4$ g (b) $34 \cdot 7$ years (c) Graph of $m = 500\ e^{-0 \cdot 02\ t}$

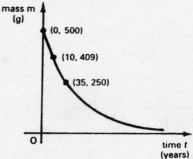

mass m (g)

$(0, 500)$

$(10, 409)$

$(35, 250)$

time t (years)

5. (a) Show $\left(\text{use } h = \frac{500}{x^2}\right)$ (b) 10 cm by 10 cm by 5 cm

6. (a) $\underline{c} = \begin{pmatrix} 2 \\ 3 \\ 4 \end{pmatrix}$ (b) $\underline{c}$ is **perpendicular** to both $\underline{a}$ and $\underline{b}$

7. (a) $x \in \{0, 70 \cdot 5, 180, 289 \cdot 5, 360\}$ (b) $f(x) = 2 \sin x; g(x) = 3 \sin 2x$

    (c) A = $(70 \cdot 5, 1 \cdot 89)$, B = $(289 \cdot 5, -1 \cdot 89)$ (d) $70 \cdot 5 < x < 180$ and $289 \cdot 5 < x < 360$

8. (a) Proof (b) Proof $\left(\text{use } \tan a° = \frac{\sin a°}{\cos a°}\right)$ (c) $8 \cdot 9$ miles

9. (a) $y = \frac{4}{3}x - 50$ (b) $x^2 + y^2 = 900$ (c) Proof; P = $(24, -18)$

10. (a) $\frac{4}{3}\ un^2$ (b) $p \doteq 2 \cdot 3$ (c) Proof; $q = 2 \cdot 3$

REVISED HIGHER ANSWERS — 1993

PAPER I

1. *(a)* $\begin{pmatrix} 2 \\ 1 \\ -2 \end{pmatrix}$ *(b)* 3 units

2. *(a)* $3y = 2x + 13$ *(b)* Proof $\{(-5, 1)$ satisfies equation (a)$\}$

3. $k \leqslant 2$ 4. $x = -1$ or 2

5. *(a)* $(10, 6)$ *(b)* $(x - 10)^2 + (y - 6)^2 = 4$ 6. Proof

7. $p = -17$ and $x = 2$ or $\frac{5}{2}$ are the other roots

8.

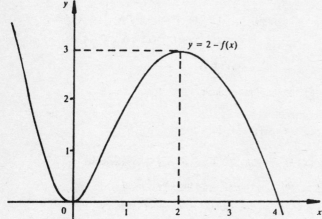

$y = 2 - f(x)$

9. $\dfrac{2}{\sqrt{x}} - 6 \sin 2x$

10. *(a)* $a = 63 \cdot 4$; $b = 135$ *(b)* $71 \cdot 6°$

11. $f(x) = x^2 - 3x + 4$

12. *(a)* 0   *(b)* $\underline{b} + \underline{c}$ is **perpendicular** to $\underline{a}$

13. *(a)* $k(x) = 5 - 4x$
    (b) $h(k(x)) = x$
    (c) $h(x) = k^{-1}(x)$ {inverses}

14.

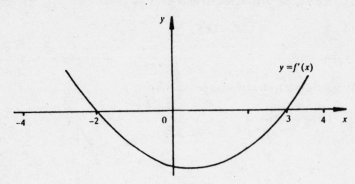

15. *(a)* $a = 5; k = 0.462$ *(b)* $y = 5e^{0.462(x-3)}$

16. $\frac{2}{9}(1 + 3x)^{3/2} + C; \frac{14}{9}$

17. $f(a) = -6\cos^2 a - \cos a + 6; a = 1.231$ or $2.094 \left\{\frac{2\pi}{3}\right\}$

18. Proof $\{g^2 + f^2 - c < 0 \text{ (i.e. } r \text{ is not } \textbf{real})\}$ **19** *(a)* Proof   *(b)* $x - \frac{1}{2}\cos 2x + C$

20. $q = p^2; q = 25$ **21.** Proof $\{f'(x) < 0; x \neq -1\}$

PAPER II

1. Minimum T.P. at $\left(-\frac{1}{2}, -\frac{27}{16}\right)$ Point of Inflexion at $(1, 0)$

2. *(a)* $A(-12, 0); B(12, 0)$ *(b)* £816

3. *(a)* $x + 2y = 2$ *(b)* $(x - 4)^2 + (y + 1)^2 = 45$

4. *(a)* 1 and 7   *(b)* $t = 1$

5. *(a)* $A(1, 0, 0); B(3, 2, 0); C(3, 0, -2)$ *(b)* $2\sqrt{3}$ units^2 *(c)* 4.7% **decrease**

6. *(a)* $A\left(\frac{7\pi}{12}, 0\right); B\left(\frac{11\pi}{12}, 0\right)$ *(b)* 3   *(c)* $C\left(\frac{\pi}{2}, 1\right)$ lies **on** the line l

7. *(a)* $B(-2, 5)$ *(b)* $m_{\text{road}} = m_{\text{circuit}}$ at B or double route implies tangency

8. *(a)* 13.05   *(b)* 3 doses   *(c)* Proof $\{u_{n+1} = 0.854 u_n + 25\}$ *(d)* No as limit 52.3 < 55

9. *(a)* Proof {various methods e.g. Cosine Rule}   *(b)* $8\sqrt{5}\sin(x - 63.4)°$ *(c)* $\theta = 114.9$

10. *(a)* $\log_e I = -\frac{4}{5}\log_e t + 4$ *(b)* Proof; $k = 54.6, r = -\frac{4}{5}$

11. *(a)* Proof {use Pythagoras}   *(b)* Proof (use differentiation); £127 million, 109 km

REVISED HIGHER ANSWERS — 1994

PAPER I

1. $\frac{3}{4}x^4 + 2x^2 + c$ 2. $k = 3$

3. $D(-11, 2, 3)$ 4. Proof {show that $\vec{RS} = 3\vec{ST}$}

5. Centre $(1, 1)$, proof {show that $m_1 \times m_2 = -1$}

6. Exact value $= 0$

7. $\underline{u} + \underline{v} = \begin{pmatrix} -2 \\ 8 \\ 2 \end{pmatrix}$; $\underline{u} - \underline{v} = \begin{pmatrix} -4 \\ -2 \\ 4 \end{pmatrix}$; Proof {use scalar product}

8. (a) $A(6, 6)$; $B(-2, -2)$ (b) $(x - 2)^2 + (y - 2)^2 = 32$ OR $x^2 + y^2 - 4x - 4y - 24 = 0$

9. (a) $u_2 = 4 \cdot 7$ (b) $n = 7$ (c) Limit $= 20$

10. $\frac{-3}{x^4} - 3\sin 3x$

11. $f(x) = (x + 4)^2 + 2$; min T.P. at $(-4, 2)$

12. (a) $a = 2$; $b = 1$ and $c = 2$ (b) $x = \frac{\pi}{12}, \frac{5\pi}{12}$

13. (a) $\frac{24}{25}$ (b) $\frac{336}{625}$

14. $m = 4$; angle $\doteq 31°$ 15. $x = 60°, 300°$

16. (a)

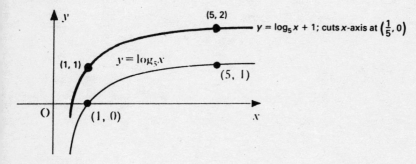

(b)

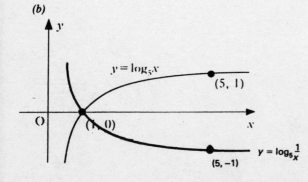

91

17. $3 \sin^2 x \cos x; \frac{1}{3} \sin^3 x + c$

18. $S(1, 0, 0)$ and $T(7, 0, 0)$

19. $f(f(x)) = \frac{x}{1 - 2x} \left\{ x \neq \frac{1}{2} \right\}$

20. $x \doteq 3 \cdot 402$; $P(3 \cdot 402, 42)$

PAPER II

1. *(a)* $a = 6$; point $(0, 6)$ *(b)* $(-3, 0)$ and $\left(\frac{1}{2}, 0 \right)$

2. *(a)* $2x + y = 10$ *(b)* $D(4, 2)$ *(c)* Area $= 5$ units^2

3. *(a)* $B(6, 4, 2)$; $C(4, 3, 4)$; $D(6, 2, 2)$ *(b)* Proof $\left\{ \text{Use } \dfrac{\vec{OA} + \vec{OD}}{2} \right\}$

   (c) $\angle AOB = 44 \cdot 4°$ *(d)* $\angle OAB = 67 \cdot 8°$

4. *(a)* (i) Distance $\doteq 78 \cdot 3$ units $\{5\sqrt{245}\}$ (ii) Clearance $\doteq 13 \cdot 3$ cm
   (b) (i) $m_{PB} = 1$ (ii) $x + y = 10$; $A(8, 2)$ and $C(6, 4)$

5. *(a)* $\sqrt{10} \sin (x - 18 \cdot 4)°$ *(b)* $x = 63 \cdot 4, 153 \cdot 4$ *(c)* $0 \leqslant x \leqslant 63 \cdot 4$ and $153 \cdot 4 \leqslant x \leqslant 180$

6. *(a)* Proof $\{$show $f(0) < 0$ and $f(0 \cdot 5) > 0\}$ *(b)* $x \doteq 0 \cdot 31$

7. *(a)* Proof $\{$area of rectangle $-$ area of $3 \triangle s\}$
   (b) Greatest area $= 24$ units^2; Least area $= 16$ units^2

8. *(a)* Proof; $f(x - 1) = 4x^2 - 11x + 12$; Proof *(b)* $4x + 7$ *(c)* $6x + 5 \{h'(x)\}$

9. *(a)* $p = \frac{-1}{2}(q + 2)$ *(b)* $p = -3$; $y = x^2 - 3x + 4$

   (c) Discriminant $= -7$ $\{$No real roots$\}$. The curve doesn't cross the x-axis

10. Area $= \frac{208}{3}$ m^2; Volume $= 4160$ m^3

PAPER I

1. 4

2. *(a)* Proof   *(b)* $f(x) = (x + 4)(2x + 1)(x - 3)$

3. $2x^3 - \dfrac{x^2}{2} + \sin x + C$

4. $k = 3$

5. $4x + 3y = 1$

6. YES; Belligerent will

7. $-\dfrac{8}{x^3} + \dfrac{3}{2}\sqrt{x}$

8. $x = \dfrac{\pi}{6}, \dfrac{5\pi}{6}, \dfrac{7\pi}{6}, \dfrac{11\pi}{6}$

9. $(x - 1)^2 + (y - 2)^2 = 18$

10. *(a)* $b = -10$ *(b)* Proof {Hint; show $f'(-2) > 0$}

11. *(a)* $\dfrac{x^2 - 2}{x^2 - 4}$ *(b)* $x \neq \pm 2$

12. $\dfrac{3\sqrt{11}}{10}$

13. $k = \sqrt{29}; x = 68\cdot 2$

14. $-\dfrac{1}{3}$

15. $x = 60, 180, 300$

16. 9

17. *(a)*   *(b)* $f'(x) = -\sin x$

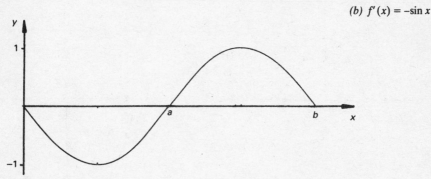

18. *(a)* $k = 0\cdot 035$ *(b)* $19\cdot 8$ minutes

19. $a = 3; b = 2$

20. $k = -5$ or 3

21. *(a)* 20 m s^{-1} *(b)* 0; Ball is stationary at a point where $h = 20$. This is its maximum height.

PAPER II

1. *(a)* Proof {Hint: show AB = AC}   *(b)* (i) $\text{H}\left(4, \dfrac{7}{2}\right)$ (ii) Proof

2. *(a)* Proof {Hint: use $y - f(2) = f'(2)(x - 2)$}   *(b)* $(-2, -18)$

3. Pestkill

4. *(a)* (i) $a = 2; b = 3$ (ii) $d = 3; c = 3$
   (b) $h(x) = \sqrt{13}\sin(3x + 56\cdot 3)°; p = 3; q = \sqrt{13}; r = 56\cdot 3$

5. *(a)* $\angle\text{PQR} = 60°$ *(b)* (i) $\text{T}\left(\dfrac{7}{3}, \dfrac{10}{3}, \dfrac{4}{3}\right)$ (ii) Proof
   (c) (i) Proof   (ii) T lies in the plane of P, Q and R

6. $x = 3; y = -1; z = 1$

7. $a = -\dfrac{1}{3}; b = \dfrac{1}{3}; c = 3$

8. $(x - 4)^2 + (y + 3)^2 = 225$

9. *(a)* Proof   *(b)* $-\dfrac{3}{4}\cos x + \dfrac{1}{12}\cos 3x + C$

10. *(a)* 48 units^2 *(b)* Proof $\left\{\text{Hint: use } \int_0^p \left(2x - \dfrac{x^2}{6}\right) dx = 24\right\}$
    (c) (i) Proof   (ii) $p \doteqdot -4\cdot 4$ or $16\cdot 4$ (iii) $-4\cdot 4 < 0$ and $16\cdot 4 > 12$

11. *(a)* Proof   *(b)* Proof {Hint: solve $\dfrac{dT}{dx} = 0$}   *(c)* $x = 4; T = 10$

93

PAPER I

1. $3x + 2y = 17$

2. $a = 2.5$

3. Max T.P. $\left(\frac{5\pi}{6}, 3\right)$

4. $2y + x = 10$

5. 9

6. B divides AC in the ratio $2 : 3$

7. $x(x-1)(x^2 + x + 1)$

8. (i)

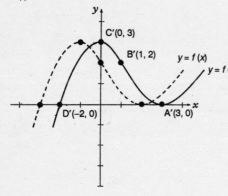

(ii)

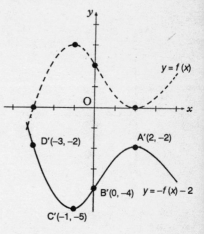

9. $\frac{5}{16}$

10. $x = 0, 120, 180, 240$

11. *(a)* This sequence has a limit since $0.3 < 1$ {i.e. the coefficient of U_n}   *(b)* $\frac{50}{7}$

12. $P\left(\frac{\pi}{6}, 0.5\right) ; Q\left(\frac{\pi}{2}, 0.5\right)$

13. $\frac{dy}{dx} = \frac{-\sin x}{2\sqrt{1 + \cos x}}$

14. They are **NOT** concurrent

15. Proof {Hint: use $\cos(x + y) = \cos x° \cos y° - \sin x° \sin y°$}

16. $x < 2$ and $x > 3$

17. $4(x + 1)^2 - 9$

18. $\sin x° = \frac{4\sqrt{5}}{9}$

19. *(a)* $k = 0.019$ *(b)* $18.75\,°C$

20. $(x - 3)^2 + (y - 4)^2 = 25$ or $x^2 + y^2 - 6x - 8y = 0$

1. *(a)* $(0, 3)$ and $(3, -24)$ *(b)* Point of inflection $(0, 3)$; Min T.P. $(3, -24)$

2. *(a)* Proof {Hint: show $m_{AB} \times m_{BC} = -1$}
   (b) (i) Eqn_{AD} $3y = x - 6$; Eqn_{BE} $3y = -4x - 1$ (ii) Hence $M\left(1, \frac{-5}{3}\right)$

3. *(a)* $Q(2, 2, 9)$; $R(21, 3, 12)$ *(b)* $Q\hat{P}R = 83 \cdot 4°$

4. *(a)* (i) $g(f(x)) = 4x^2 + 4x + 1 + k$ (ii) $f(g(x)) = 2x^2 + 2k + 1$
   (b) (i) Proof   (ii) When $k = 6$ the roots are real, distinct and rational   (iii) $k = -2$ for equal roots

5. Area $= 1 - \frac{\sqrt{3}}{4}$ units^2 {$\approx 0 \cdot 567$}

6. *(a)* Eqn **l** is $y = 2x$; Eqn **p** is $y = 2x^2$; Eqn **c** is $x^2 + y^2 = 5$
   (b) Equ **l**′ is $y = 2x - 4$; Eqn **p**′ is $y = -2x^2$; Eqn **c**′ is $(x - 2)^2 + y^2 = 5$
   (c) (i) Proof   (ii) 2 points $(1, -2)$ and $(-2, -8)$

7. *(a)* $f(x) = \sqrt{13} \cos (x - 56 \cdot 3)°$ *(b)* $x = 138 \cdot 3$ and $334 \cdot 3$

8. *(a)* $\text{Eqn}_{\text{tangent}}$ $5x + y + 3 = 0$; $B(-1, 2)$ *(b)* Area $= \frac{4}{3}$ units^2

9. *(a)* $y = 3x + 0 \cdot 7$ *(b)* $a \approx 2 \cdot 0$; $b = 3$

10. *(a)* Proof {Hint: use $m_1 \times m_2 = -1$}
    (b) (i) $2y = 3x + 23$ (ii) $x^2 + \left(y - \frac{23}{2}\right)^2 = \frac{117}{4}$ or $x^2 + y^2 - 23y + 103 = 0$

11. *(a)* (i) $h = 5 - x - \frac{\pi}{2}x$ (ii) Proof   *(b)* $x \approx 1 \cdot 15$; $h \approx 2 \cdot 05$

Topic	1990 Paper I	1990 Paper II	1991 Paper I	1991 Paper II	1992 Paper I	1992 Paper II	1993 Paper I	1993 Paper II	1994 Paper I	1994 Paper II	1995 Paper I	1995 Paper II	1996 Paper I	1996 Paper II
The Straight Line	20	2	1,2		2		2,10			2	5,6	1	1,14	2
Differentiation 1 Basics, Max, Min, Graph	2,11,16	1	5	1,10	1,19	1,5	4	1,7	2,14	7	7,10,14, 21	2(a),7	9,16	1,8(a), 11
Quadratic Theory	18		18		17		3		11		20		2,17	4(b)
Trigonometry Reminders and Radians	10,15	10		3(a)		8	17		12		8	4(a)	3,12	
Sequences		3	11	9		3		8	9	6		3	11	
Functions and Graphs	6	6(a)(c)	9,14,15, 19		6,10		8,13,14		19		11		8	4(a),6
Compound Angle Formulae (trig.)	9		12,20	3(b)	5,13	7	6,19(a)	6	6,13,15		12,15	9(a)	10,15,18	
Differentiation 2 Trig. & Chain Rule	19		13		11		9,21	11	10,17		17	11	13	
The Circle	3,7	8	8	2	9,16	9	5,18	3	5,8	4	9	8	4,20	10
Polynomials Remainder Theorem	1	6(b)	6		3		7		1	1	2	2(b), 10(c)	7	
Integration	8,13	7	10,16	11	4,8,14	10(a)(b)	11,19(b)	2	1	10	3	9(b),10	5	5,8(b)
Vectors	4,5,12	4	3,7,17	5	15,18	2	1,16	5	3,4,7,18	3	1,4,16	5	6	3
The Wave Function		5		8	7	10(c)		9		5	13	4(b)		7
Exponential and Log Functions	14,17		4	4,7	12	4	15,20	10	16,20		18,19		19	9
Miscellaneous Questions		9		6		6		4		8		6		

Printed by Bell & Bain Ltd., Glasgow.